Exploring Science

Program Consultants

Randy L. Bell, Ph.D.
Malcolm B. Butler, Ph.D.
Kathy Cabe Trundle, Ph.D.
Judith S. Lederman, Ph.D.
Center for the Advancement of Science in Space, Inc.

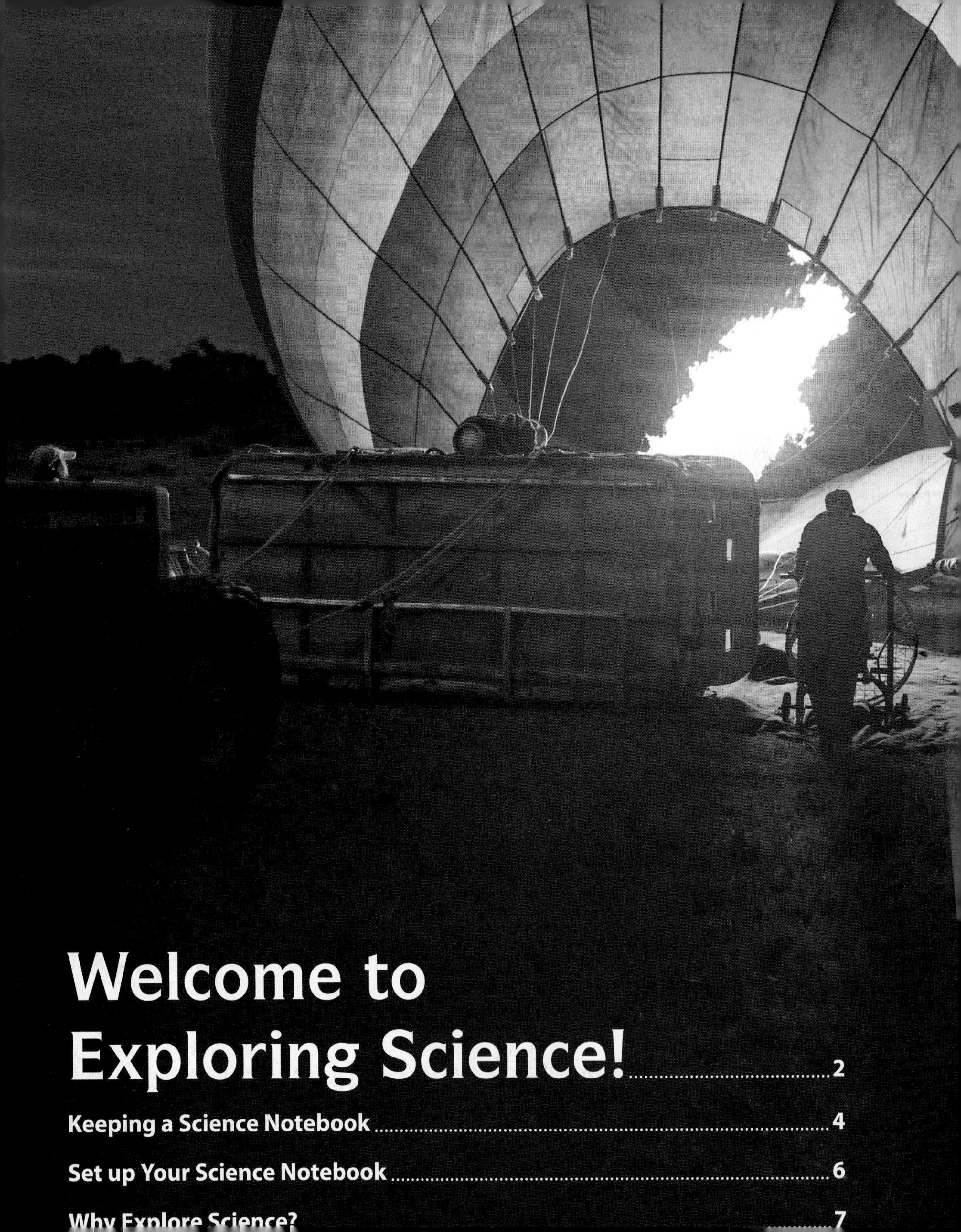

Welcome to Exploring Science!

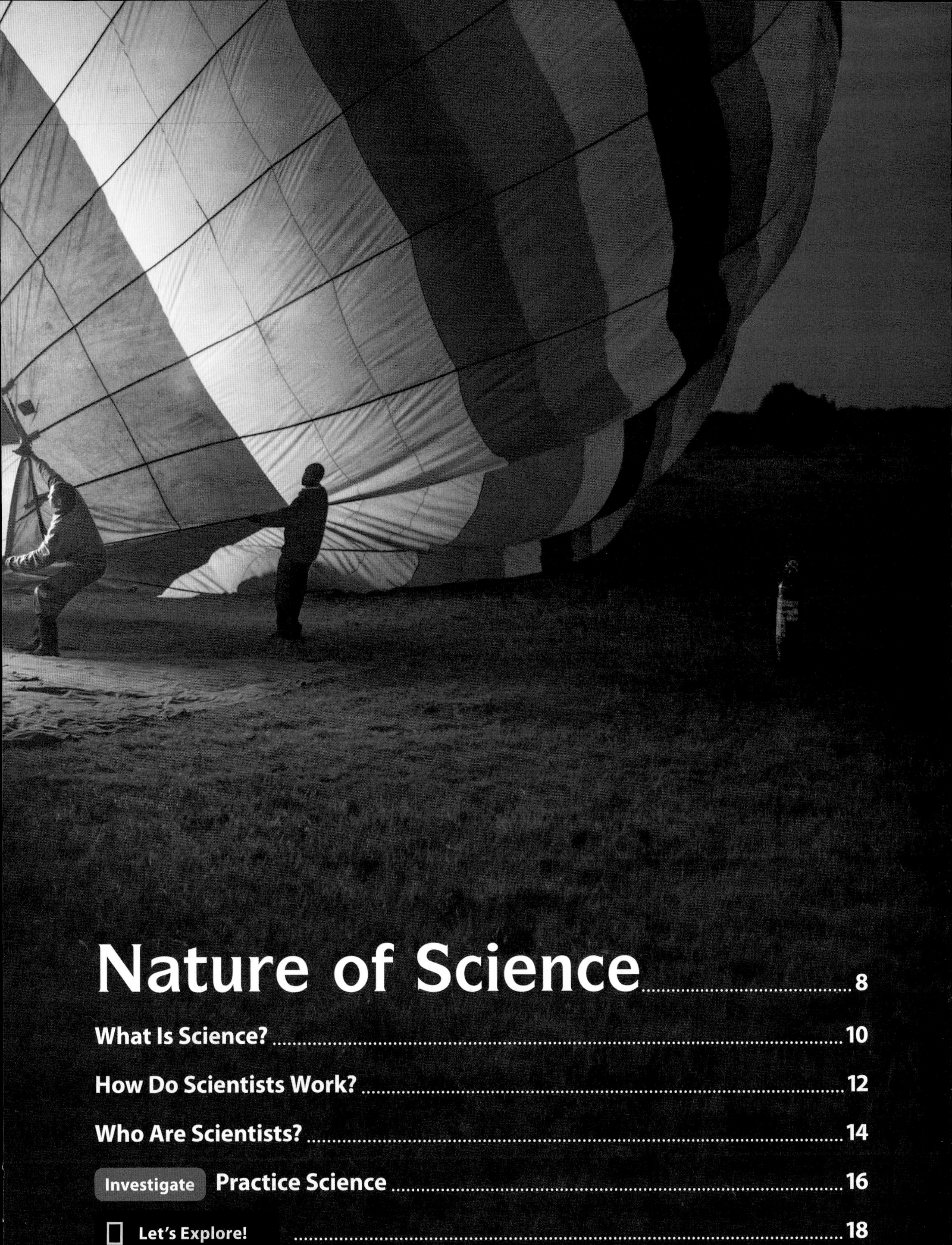

Nature of Science

Physical Science

Life Science

Earth Science

Earth's Systems

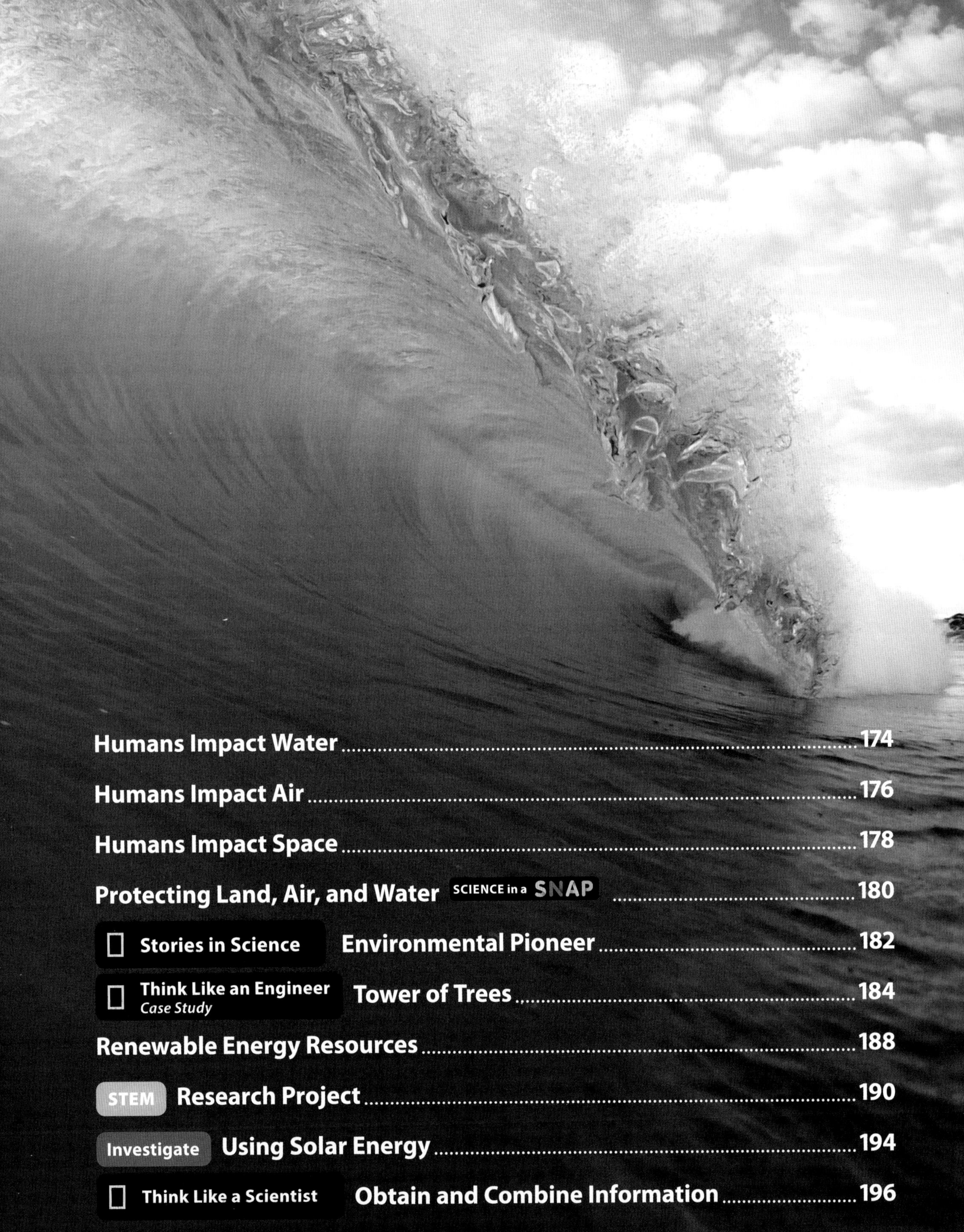

Earth Science (continued)

Space Systems: Stars and the Solar System

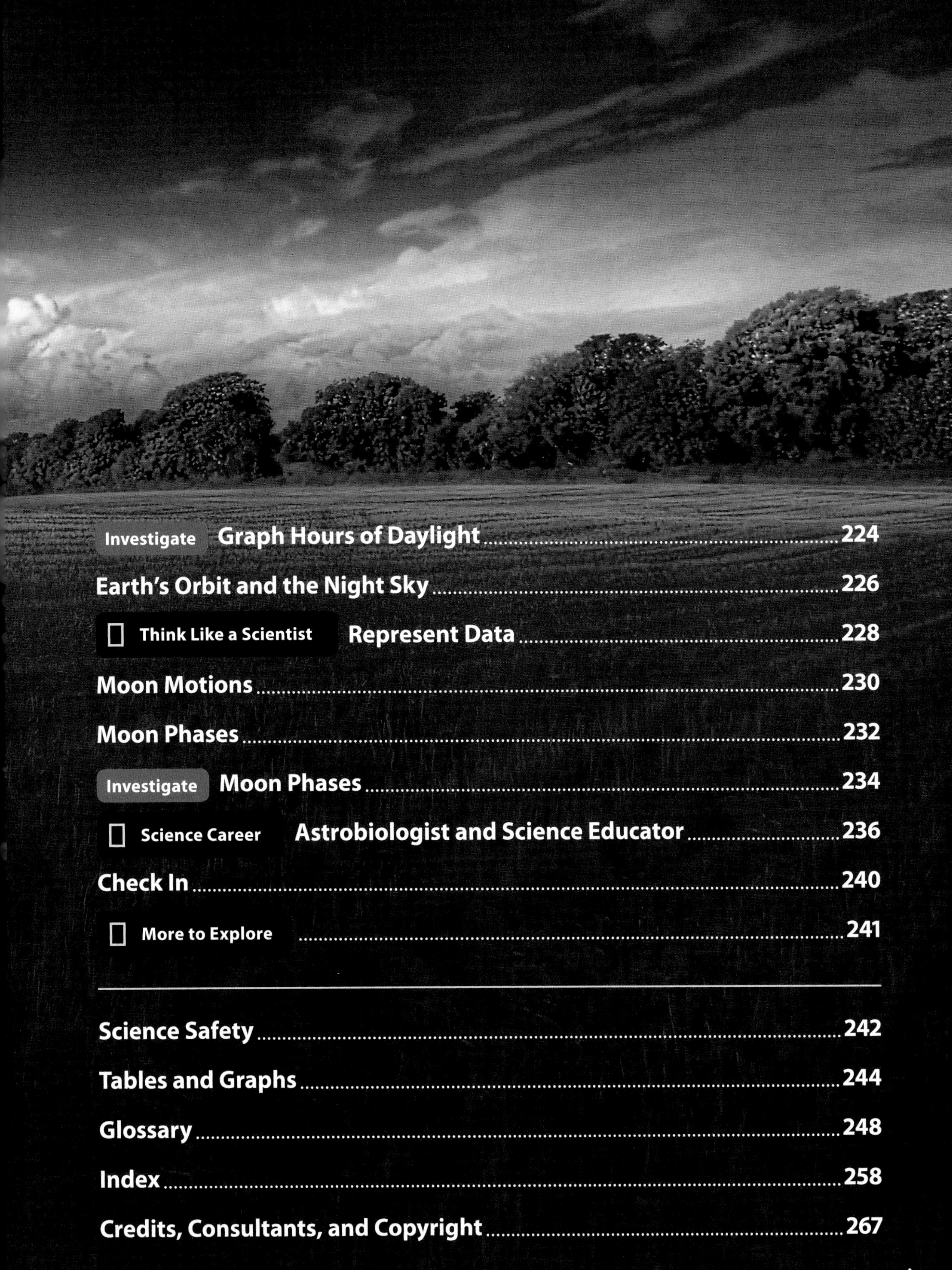

Explorer

Andrés Ruzo is a geoscientist and National Geographic Explorer. He studies Earth's heat and how it can be harnessed for power. Andrés believes that solving energy issues will also take care of other major world problems.

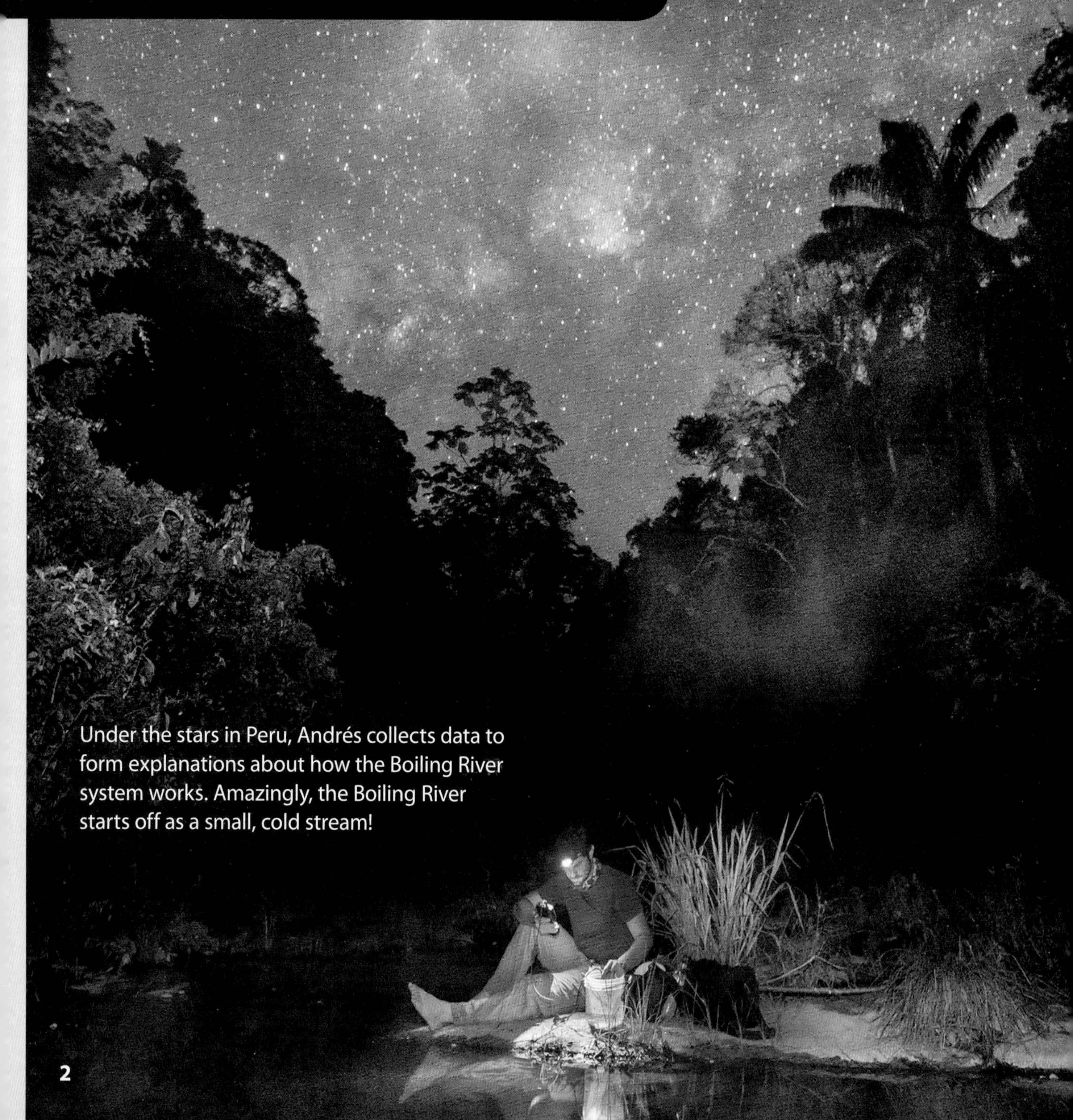

Under the stars in Peru, Andrés collects data to form explanations about how the Boiling River system works. Amazingly, the Boiling River starts off as a small, cold stream!

Welcome to Exploring Science!

¡Hola, exploradores! I am Andrés Ruzo. Together, in *Exploring Science,* let's investigate how scientists answer questions, do investigations, and make new discoveries. When you observe the world around you and draw logical, repeatable conclusions, you are doing science! Some conclusions become more refined or are replaced as knowledge advances. Some conclusions can be meaningful even after thousands of years!

I research Earth's heat and features, such as volcanoes, hot springs, and boiling rivers. The Boiling River in Peru is one of my favorite places to work. It flows hot for 6.2 kilometers (3.8 miles)! There is even a 6-meter (20-foot) scalding waterfall! Though other thermal rivers exist in the world, Peru's Boiling River is over 700 kilometers (400 miles) from the nearest active volcano. I want to understand how this system forms and what impact it has on humans, and I want to help the community protect the area for future generations.

Keeping a Science Notebook

My Science Notebook

I use a science notebook to keep a careful record of my work. I record observations, measurements, and other data. I analyze the data to make explanations and conclusions based on evidence.

In *Exploring Science,* you will use your own science notebook to keep a record of your work. You can also share what you learn with others. What will your notebook include? Look at the list and examples that follow for ideas. You and your teacher might have ideas, too. Then get ready to set up your science notebook.

- Define and illustrate science vocabulary and main ideas.
- Label drawings. Write notes to explain ideas.
- Collect photos, news stories, and other objects.
- Use tables, charts, or graphs to record and analyze data.
- Include evidence for explanations and conclusions.
- Describe how scientists and engineers answer questions and solve problems.
- Ask new questions of your own.

A science notebook is one of a scientist's most important tools.

In your science notebook, write new questions you may have.

My New Questions

1. How could I safely test to see if any liquids are good electrical conductors?
2. Which solid and liquid materials can dissolve in vegetable oil?
3. How can I test to see if the mass of vinegar and baking soda is the same before and after mixing?
4. How are different colors produced in fireworks?

Label drawings, and include captions to explain ideas.

Use tables, charts, or graphs to record observations and data in investigations.

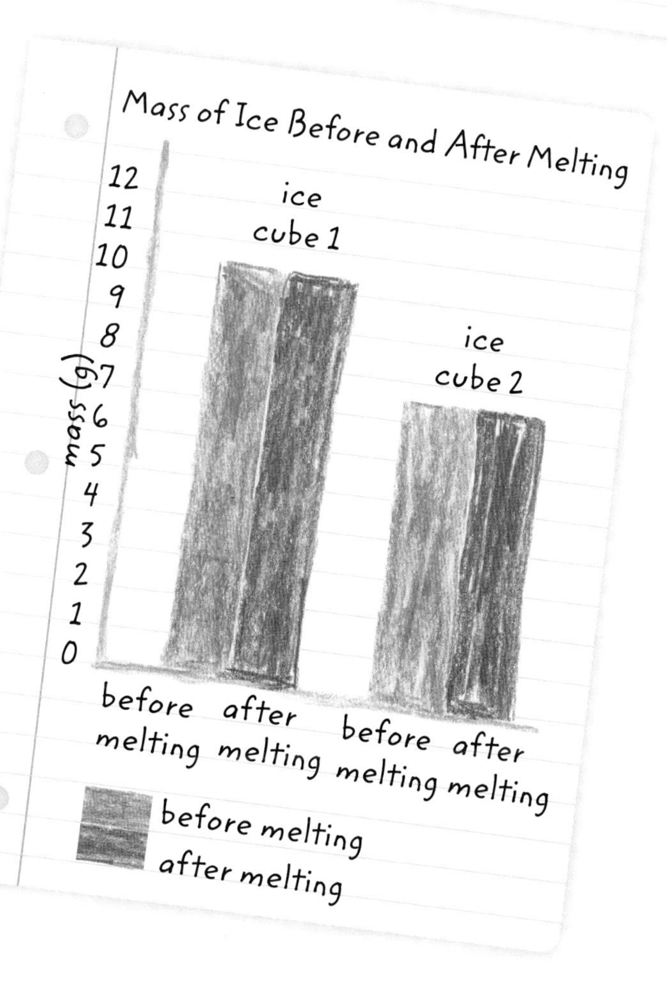

Set up Your Science Notebook

My Science Notebook

Use your science notebook every time you study science. Here are a few suggestions to make your notebook unique and easy to use. Your teacher may have more instructions. Use your own ideas, too!

- Design a cover for your notebook. Include something you like about science or something you would like to learn.
- In the front of your notebook, write "Table of Contents." Leave some blank pages. You will need to add the date, title, and page for each entry.
- Organize your notebook. You might make entries in the order you read and do investigations. Write the date for any new entry.
- Add a page number to each new page you write on.
- Keep your science notebook in a safe place to keep it in good condition. You'll want it to last, so you can see how much you have learned!

Design a cover that is all about science and you!

Make a table of contents. Write dates, titles, and page numbers as you work.

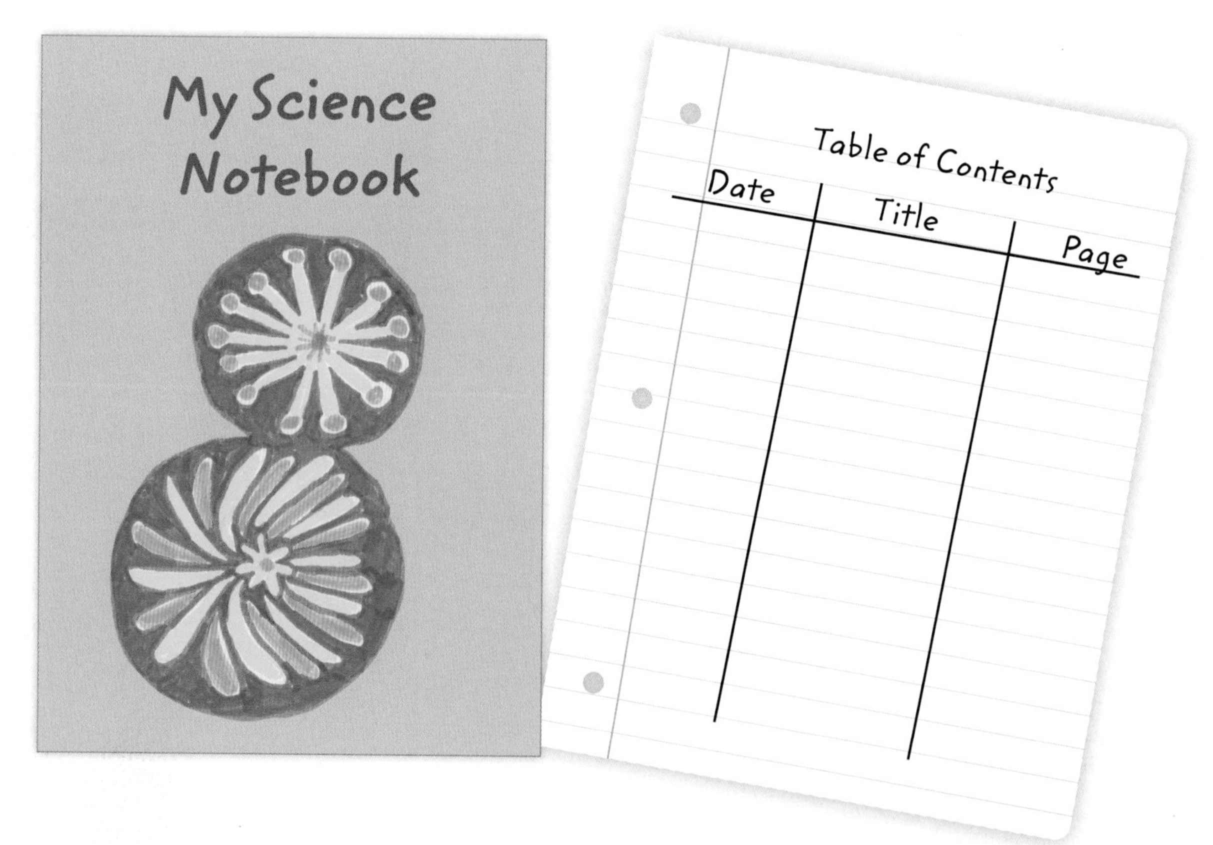

Why Explore Science?

Science is a way of learning how things work in the natural world. You act like a scientist when you ask questions, do investigations, and analyze observations and data to make conclusions. I ask questions and investigate to learn about the Boiling River. I wonder why the water in this small stretch of river so far from a volcano is so hot. I make careful observations at the Boiling River. I work along with the local people to learn.

What questions do you have about the natural world? Have you ever observed something and wondered "Why did that happen?" or "How does that work?" As you study science, observe and ask questions. Make explanations by inferring based on what you know and observe. Then you are acting like a scientist.

Next you will read about the *Nature of Science*. You will learn more about what science is and ways that scientists work. As you read, think about questions you have. Write them in your science notebook!

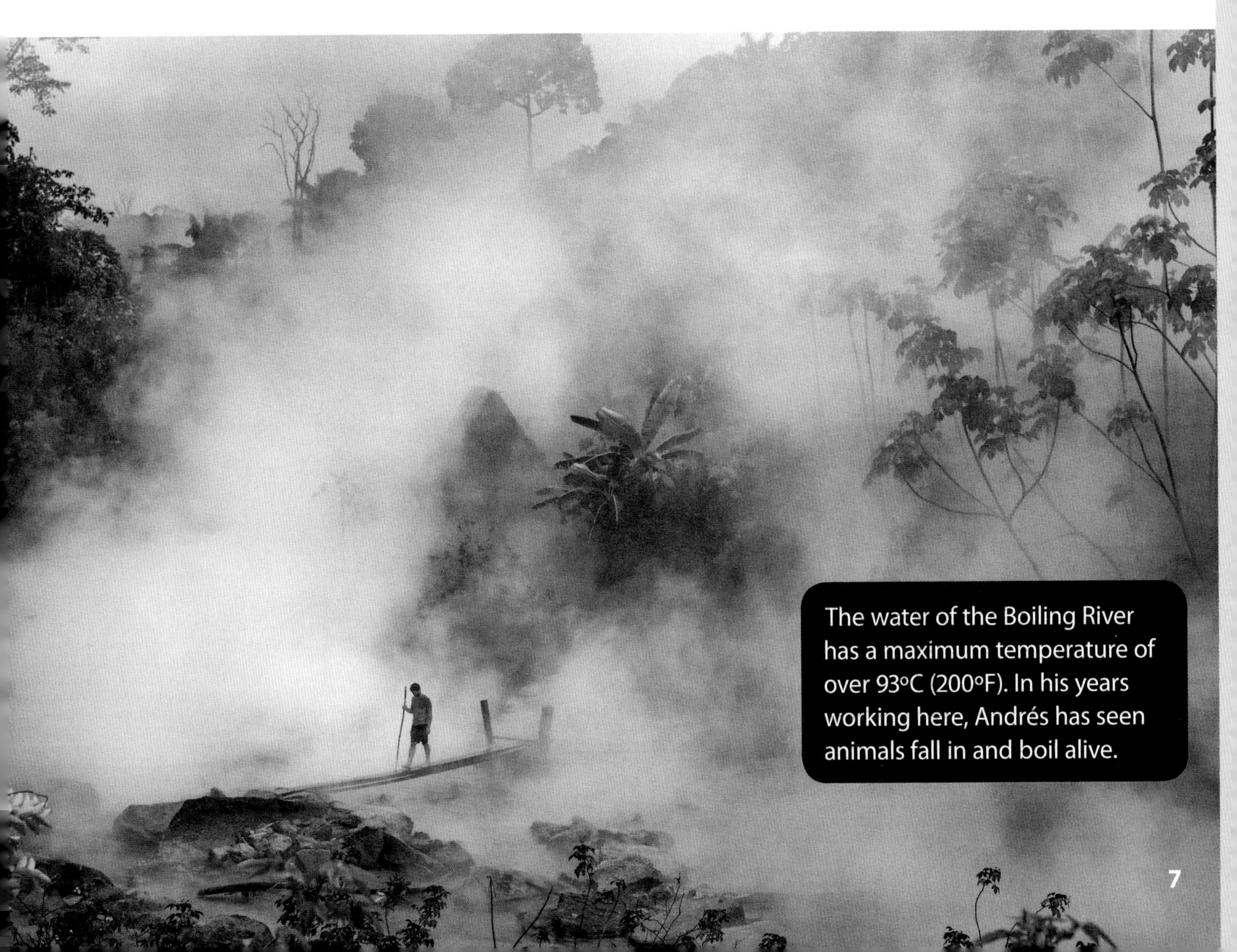

The water of the Boiling River has a maximum temperature of over 93°C (200°F). In his years working here, Andrés has seen animals fall in and boil alive.

Nature of Science

Long ago, some people hoped to find a way to turn metals such as lead into gold. They did not have a good understanding of how the many changes in matter could happen. However, they did make some new discoveries. The material shown in the painting is phosphorus, a substance that can glow when it burns.

What Is Science?

Science is a way of knowing about the natural world. This knowledge grows as we improve explanations about what we observe. To **observe** means to use your senses to collect information about our world. Observations often lead to new scientific questions to answer. The process never ends!

Look back at the photo of a painting of a laboratory in the Middle Ages. People once believed there was a way to change metals of low quality into metals of high quality, such as turning lead into gold. They tried many ways to make changes in matter. This process is called alchemy. Alchemists made observations and inferences. To **infer** is to draw conclusions based on what is already known and new observations. But alchemists' conclusions were heavily influenced by their beliefs. Science knowledge is based on evidence, not beliefs. **Evidence** is information that comes from analyzing and making sense of **data,** such as measurements and other observations. Evidence supports an idea or conclusion.

Through the work of many scientists over time, we now understand much more about changes in matter. Look at the photos. Scientists use evidence to explain these changes.

When some materials are combined, a new material can form. The clear yellow liquid is added to a different, colorless liquid. A new yellow material is formed. This is just one way matter can change.

NS Science Models, Laws, Mechanisms, and Theories Explain Natural Phenomena.
Science explanations describe the mechanisms for natural events. (5-LS2-1)
NS Scientific Knowledge Assumes an Order and Consistency in Natural Systems.
Science assumes consistent patterns in natural system. (5-PS1-2)

Have you ever seen flames of different colors? Different types of metal materials are added to the wax in the candles. When the materials are burned, new materials form. Flames of different colors are produced.

Wrap It Up!

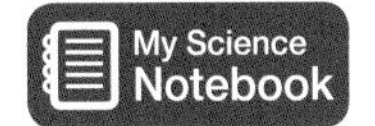

1. **Describe** Write about or draw an observation you made today. Think of a question to ask about your observation.
2. **Infer** Why do you think it is important that scientists continue to improve our knowledge of the natural world?
3. **Apply** You observe ice melt and refreeze. Make an inference. Has a new, different material formed? Explain your answer.

How Do Scientists Work?

Scientists use different methods to produce evidence. In an **investigation,** scientists ask questions, plan procedures for observations, collect data and look for patterns, and make conclusions. Scientists, such as chemists, may do investigations in a lab. Others, such as wildlife biologists, make direct observations in the field.

An **experiment** is a type of investigation. It is a fair test, or process in which scientists control variables to test a hypothesis. A **variable** is a factor in an experiment that may change. A **hypothesis** is an idea or explanation that can be tested by investigation.

Scientists also use models. A **model** is a representation of a natural process or system. For example, some scientists work with materials that are so small they can only be seen using very powerful microscopes. They often make computer models of the materials, so they can better visualize what they are making. Scientists also use models to explain how something works or to make predictions about how a material will behave under different conditions.

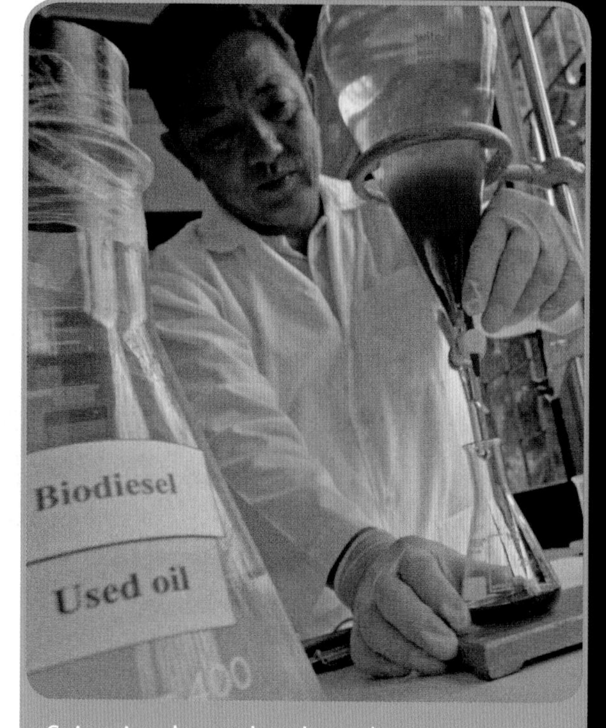

Scientists have developed ways to use cooking oil and other materials to make new kinds of fuels. Biodiesel is a fuel used by some kinds of cars and trucks. It can replace fuel made from oil, a fuel source that someday may run out.

Scientists repeat investigations to make sure they have accurate results. They also repeat investigations of other scientists to make sure they can get the same results.

NS Science Models, Laws, Mechanisms, and Theories Explain Natural Phenomena.
Science explanations describe the mechanisms for natural events. (5-LS2-1)
NS Scientific Knowledge Assumes an Order and Consistency in Natural Systems.
Science assumes consistent patterns in natural system. (5-PS1-2)

Scientists have developed an ultra-light material called aerogel. Very little heat can pass through it. The blowtorch does not melt the crayons placed on the aerogel.

Wrap It Up!

My Science Notebook

1. **Describe** What do scientists do when they perform an experiment?
2. **Describe** How can making a model be helpful to scientists?
3. **Apply** You repeat a friend's experiment and do not get the same results. What are some things you could do before you try the experiment again?

Who Are Scientists?

Scientists come from different backgrounds and cultures. They often work together in teams. For example, scientists at the U.S. Department of Energy's Argonne National Laboratory are working to make an efficient, affordable battery for electric vehicles. The team includes materials scientists, chemists, computer scientists, and physicists working in different areas of the United States.

To soak up oil spills, materials such as straw, sand, and clay are used. But these materials can be used only once. They may sink in the ocean and take the oil with them. A team of scientists have developed a new material to soak up oil spills. The material can be wrung out like a washcloth and used again.

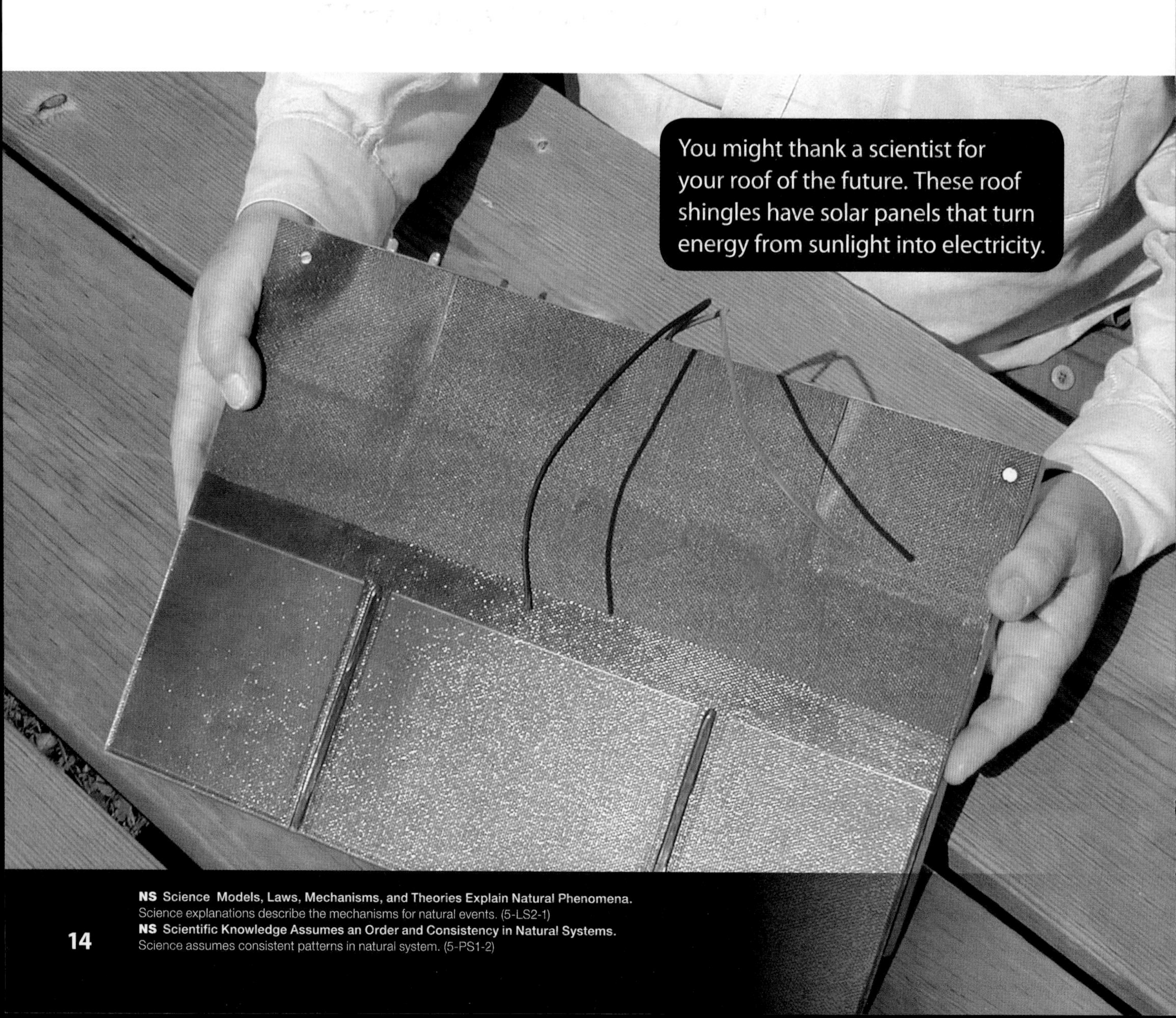

You might thank a scientist for your roof of the future. These roof shingles have solar panels that turn energy from sunlight into electricity.

NS Science Models, Laws, Mechanisms, and Theories Explain Natural Phenomena.
Science explanations describe the mechanisms for natural events. (5-LS2-1)
NS Scientific Knowledge Assumes an Order and Consistency in Natural Systems.
Science assumes consistent patterns in natural system. (5-PS1-2)

Scientists with different backgrounds bring different ideas, perspectives, and priorities to investigations. They ask different questions or propose different solutions based on their knowledge and experience.

Curiosity and creativity are important in science. Scientists always ask new questions about how or why something works as it does. Creativity involves looking at things in new ways when asking questions, planning investigations, and finding ways to answer questions. This creativity results in science that affects our lives every day. The electricity we use, the crops we grow for food, and the medicine we might need are only a few results of the work of scientists.

Now it's your turn to think like a scientist. Are you ready? Then let's go!

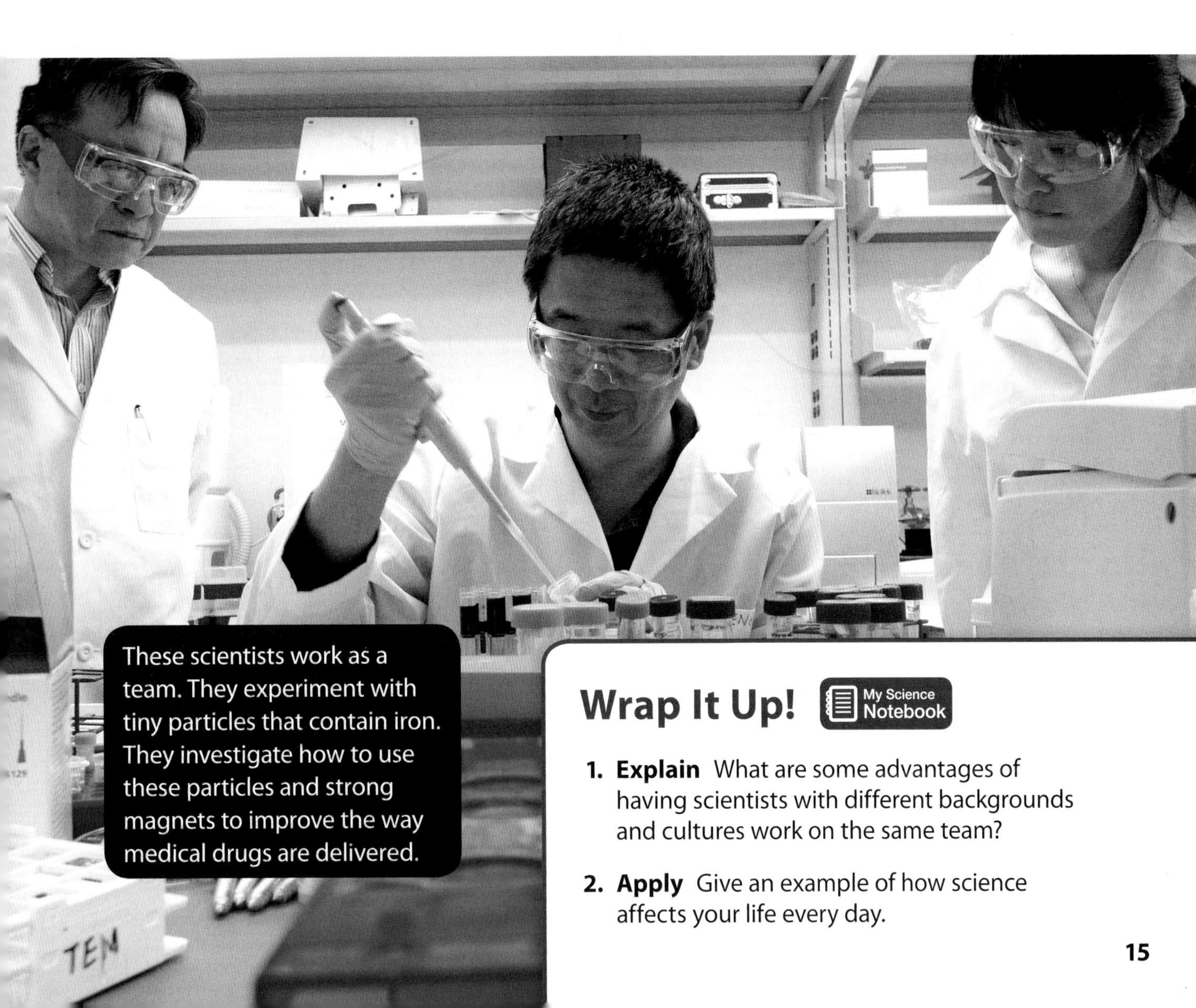

These scientists work as a team. They experiment with tiny particles that contain iron. They investigate how to use these particles and strong magnets to improve the way medical drugs are delivered.

Wrap It Up! My Science Notebook

1. **Explain** What are some advantages of having scientists with different backgrounds and cultures work on the same team?
2. **Apply** Give an example of how science affects your life every day.

Investigate

Practice Science

How can you investigate the properties of a mystery material?

Scientific investigations involve making observations, collecting and analyzing data, and making inferences based on evidence to draw conclusions. In this investigation, you will observe and infer about the properties of a mysterious material, Oobleck! First you need to know some properties of solids and liquids. A solid, such as a wooden block, has a definite shape. It can feel hard to the touch. A solid does not flow. A liquid, such as honey, is not hard. It takes the shape of a container. A liquid can flow.

Materials

newspapers

plastic cup

objects for testing the properties of Oobleck

bowl of Oobleck

resealable plastic bags

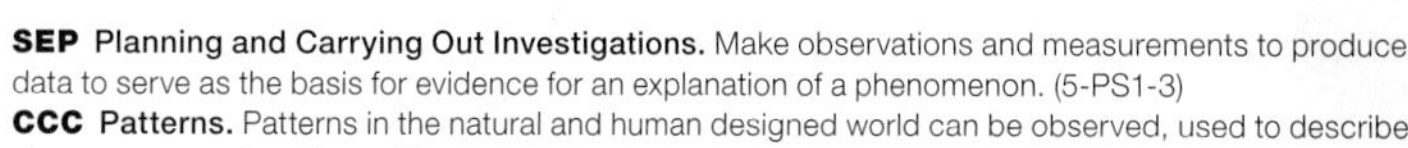

SEP **Planning and Carrying Out Investigations.** Make observations and measurements to produce data to serve as the basis for evidence for an explanation of a phenomenon. (5-PS1-3)
CCC **Patterns.** Patterns in the natural and human designed world can be observed, used to describe phenomena, and used as evidence.
NS **Science Models, Laws, Mechanisms, and Theories Explain Natural Phenomena.** Science explanations describe the mechanisms for natural events. (5-LS2-1)
NS **Scientific Knowledge Assumes an Order and Consistency in Natural Systems.** Science assumes consistent patterns in natural system. (5-PS1-2)

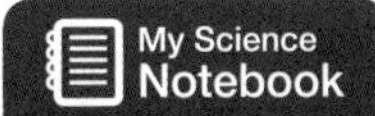

1 Observe the Oobleck. Record the properties you can observe without touching it. Now move some Oobleck into a small cup and then back into the bowl. Try moving the Oobleck quickly and then slowly. Record your observations in your science notebook.

2 Use some pressure to quickly push your fingertips into the Oobleck. Then slowly lower your fingers into it using very little pressure. Move your fingers slowly through the Oobleck. Then try moving them quickly. Record your observations.

3 Hold a handful of Oobleck in your open hand. Try squeezing it in your fist. Then try rolling it between your hands. Record your observations.

4 Look at the materials you have. Think about how the materials could be used to test the properties of Oobleck. Write questions you would like to answer. Then use the materials to try to answer your questions. Record your observations.

This liquid is magnetic. It is attracted to a magnet to create a sculpture.

Wrap It Up!

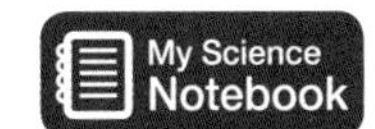

1. **Analyze** Look for patterns in your data. What properties did you observe as you moved, touched, or held the Oobleck?
2. **Infer** Under what conditions did Oobleck have properties of a solid? of a liquid? Use evidence from your investigation to explain.
3. **Conclude** Is Oobleck a solid or a liquid, or would you describe it another way? What evidence did you use to form your conclusion?
4. **Explain** How did you act as a scientist as you investigated the properties of Oobleck?

NATIONAL GEOGRAPHIC | Explorer

Andrés Ruzo Geoscientist
National Geographic Explorer

Let's Explore!

Every day, new advances in science help scientists and engineers give us a better understanding of our world and solve important problems. Science builds upon itself. More observations and new approaches to problems result in new discoveries and more focused questions. One of my favorite quotes about science is "Science is about asking better questions."

Here are some questions you might answer in *Physical Science*:

- If you melt an ice cube, does the liquid water weigh as much as when it was solid ice?
- How does the plastic coating on an electrical wire keep you safe?
- Why can't you unburn toast?
- What causes fireworks to explode, sparkle, and glow?
- If you are given sugar, cornstarch, baking soda, and baking powder, how can you determine which is which?

As you read, think of questions of your own. Then let's check in later to compare notes!

Include notes to explain science concepts.

Cause and Effect

Cause: The nail is an electrical conductor that completes the circuit when it touches the bulb holder.

Effect: Bulb lights up.

Use your science notebook to record ideas you think are important.

Electrical Conductors and Insulators

Plastic is an electrical insulator. It is used on electrical cords to prevent shocks. If the insulator is worn, the cord could be dangerous!

Copper wire is an electrical conductor. If you touch the wire, you could get a shock. Replace this cord!

You can use your notebook to reflect on what you've done and learned.

1. I thought like a scientist when I did research to obtain information on ways that communities protect Earth's resources. I looked at the problems and solutions.
2. I thought like an engineer when I read about the Tower of Trees. I thought of some ideas to solve the problem of too few trees in a city. Then I read about the solution of planting trees on buildings.
3. I learned the difference between a physical change and a chemical change. I drew examples of each in my notebook.

Physical Science

Structure and Properties of Matter

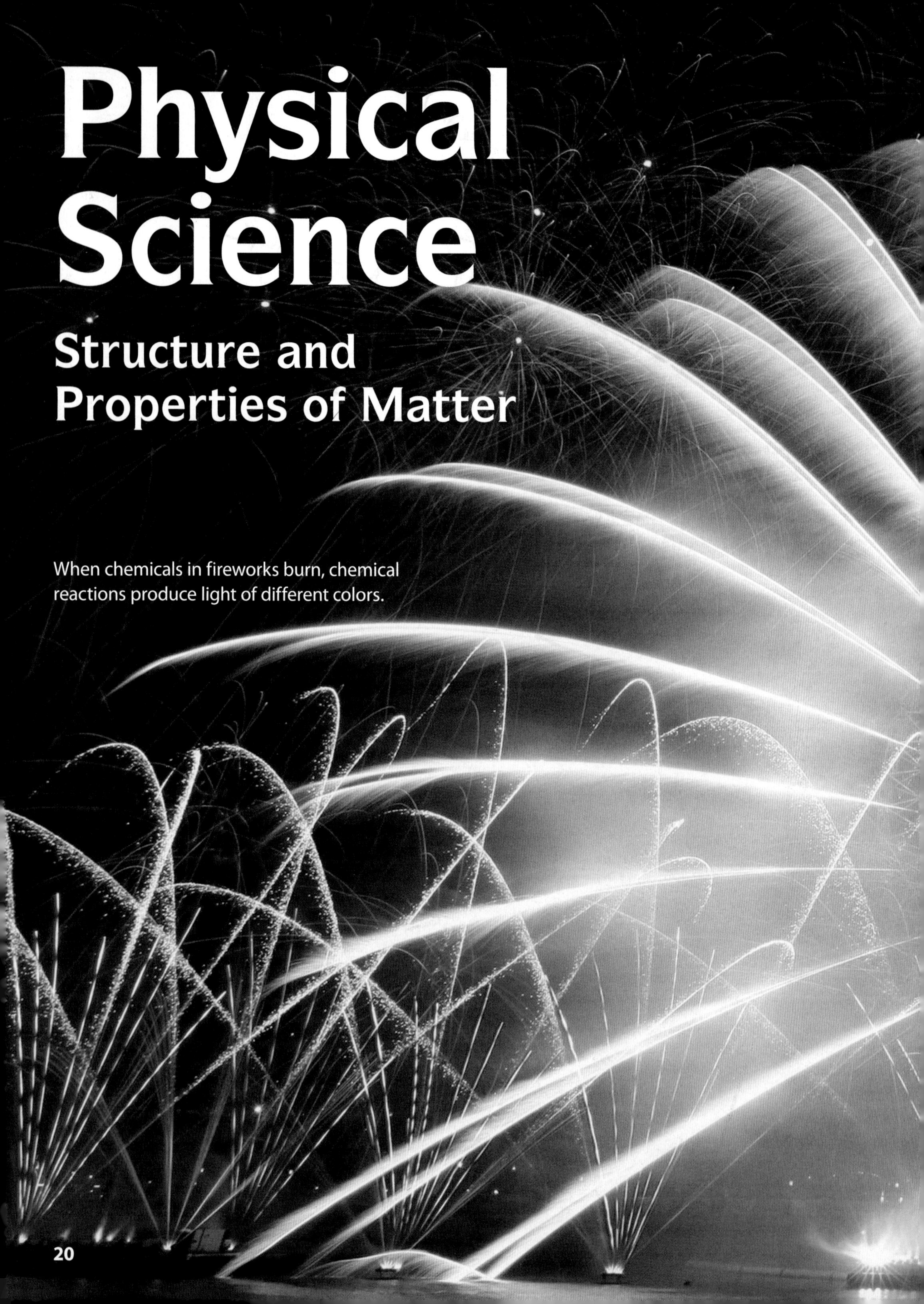

When chemicals in fireworks burn, chemical reactions produce light of different colors.

Matter

Can you identify the matter in this picture? Here's a hint—everything you see in the picture is made of matter. The sandcastle is made of matter and so is the water. Even the air is made of matter. **Matter** is anything that has mass and takes up space.

Mass is the amount of "stuff" in an object. Imagine holding two balls of the same size. One is made of foam, and the other is made of the metal lead.

This sandcastle is matter. It takes up space on the beach. If you could put it on a scale, you would see that it also has mass.

DCI PS1.A: Structure and Properties of Matter. Matter of any type can be subdivided into particles that are too small to see, but even then the matter still exists and can be detected by other means. A model shows that gases are made from matter particles that are too small to see and are moving freely around in space can explain many observations, including the inflation and shape of a balloon and the effects of air on larger particles or objects. (5-PS1-1)
CCC Scale, Proportion, and Quantity. Natural objects exist from the very small to the immensely large. (5-ESS1-1), (5-PS1-1)

Which would feel heavier? The lead ball, of course! That's because the metal ball has more matter in it, and so, more mass.

Matter of any type can be broken down into smaller parts. The grains of sand in the sandcastle came from much larger rocks that were broken down over time. They may have come from huge slabs of rock such as those you see to the right in the photo. And each grain of sand can be broken down into even smaller particles, which are too small to see. But even these tiny particles are matter.

The sandcastle is made up of thousands of grains of sand.

Even though this grain of sand has been magnified 100 times, you still can't see the particles that make it up. They are too small to see.

Wrap It Up! My Science Notebook

1. **Define** What is matter?
2. **Infer** Do sand particles have mass? Explain.
3. **Compare and Contrast** How are the slabs of rock alike and different from the tiny particles that make up a grain of sand?

States of Matter

Matter can be classified by its state. Solids, liquids, and gases are all physical **states of matter.** Each state has specific characteristics. For example, **solids** have a definite shape. A brick is a solid.

Liquids, such as milk in a bottle, take the shape of their containers. Liquids do not necessarily fill a container completely.

If you have ever seen a balloon floating in the air, you have seen an object filled with gas. **Gases** have no definite shape. Gases spread out to completely fill a closed container, such as closed plastic bottle. The gas particles may be too small to see, but you can tell they are there. If you squeeze an open plastic bottle, gas is forced out and the bottle collapses. If you squeeze a closed plastic bottle, the trapped particles of air push back.

Liquid The ocean is a liquid. Particles of a liquid are farther apart. The particles move more freely than the particles of a solid.

Solid This sailboard is a solid. The particles that make up a solid are close together and vibrate, or "jiggle," in place.

DCI PS1.A: Structure and Properties of Matter. Matter of any type can be subdivided into particles that are too small to see, but even then the matter still exists and can be detected by other means. A model shows that gases are made from matter particles that are too small to see and are moving freely around in space can explain many observations, including the inflation and shape of a balloon and the effects of air on larger particles or objects. (5-PS1-1)

CCC Scale, Proportion, and Quantity. Natural objects exist from the very small to the immensely large. (5-ESS1-1), (5-PS1-1)

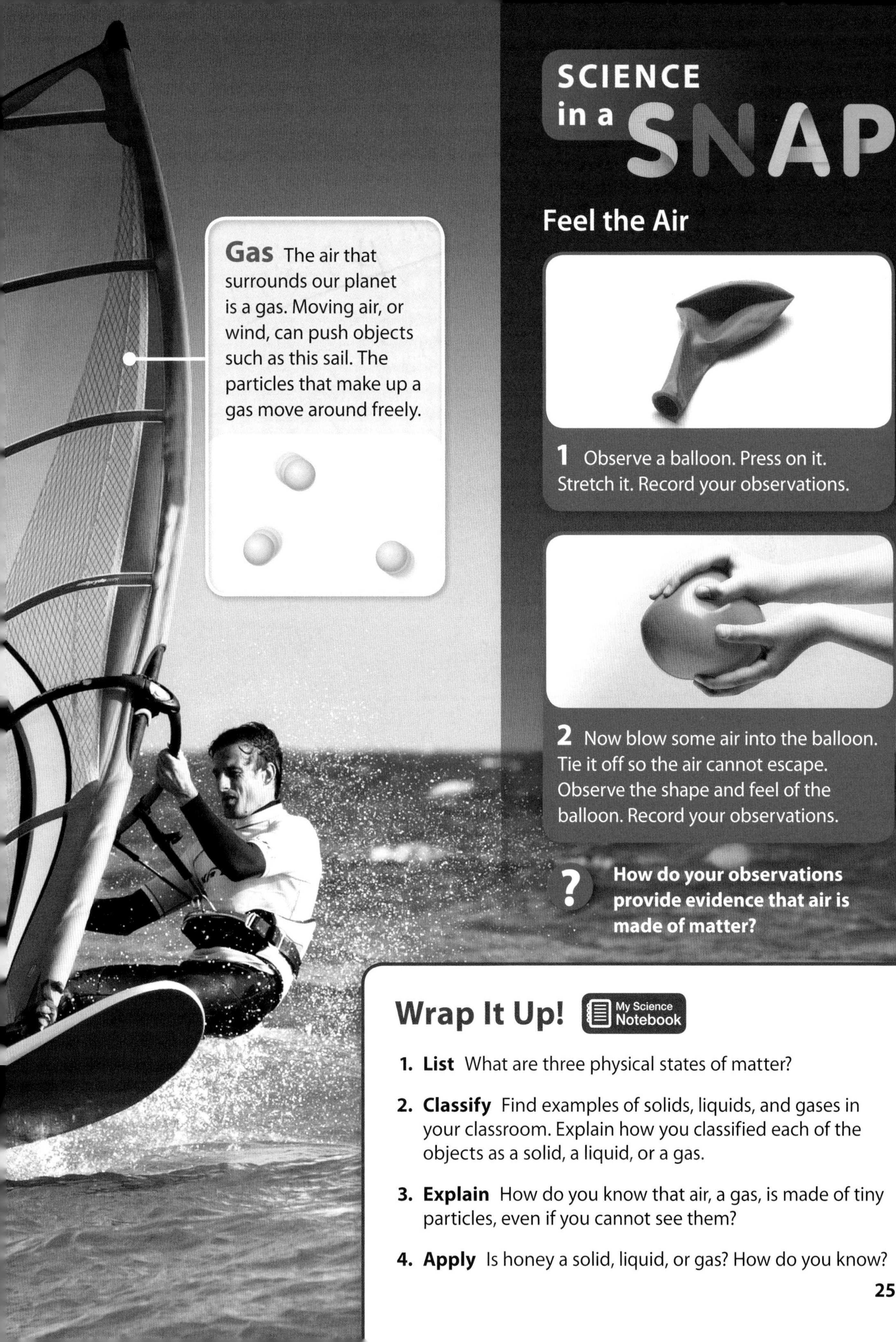

Gas The air that surrounds our planet is a gas. Moving air, or wind, can push objects such as this sail. The particles that make up a gas move around freely.

SCIENCE in a SNAP

Feel the Air

1 Observe a balloon. Press on it. Stretch it. Record your observations.

2 Now blow some air into the balloon. Tie it off so the air cannot escape. Observe the shape and feel of the balloon. Record your observations.

? **How do your observations provide evidence that air is made of matter?**

Wrap It Up! My Science Notebook

1. **List** What are three physical states of matter?
2. **Classify** Find examples of solids, liquids, and gases in your classroom. Explain how you classified each of the objects as a solid, a liquid, or a gas.
3. **Explain** How do you know that air, a gas, is made of tiny particles, even if you cannot see them?
4. **Apply** Is honey a solid, liquid, or gas? How do you know?

Investigate

Matter

How can you detect materials that have dissolved in water?

As salt is stirred into water, it seems to disappear. But the salt is not gone—it dissolves. When a solid **dissolves** in a liquid, the tiny particles that make it up become evenly mixed into the liquid. How do you know the salt is still there? If the liquid **evaporates,** or changes state from a liquid into a gas, the solid is left behind. In this investigation, you will use evaporation to separate salt from salt water.

Materials

salt	water in a cup	spoon
dropper	black construction paper	hand lens

DCI PS1.A: Structure and Properties of Matter. Matter of any type can be subdivided into particles that are too small to see, but even then the matter still exists and can be detected by other means. A model shows that gases are made from matter particles that are too small to see and are moving freely around in space can explain many observations, including the inflation and shape of a balloon and the effects of air on larger particles or objects. (5-PS1-1)
CCC Scale, Proportion, and Quantity. Natural objects exist from the very small to the immensely large. (5-ESS1-1), (5-PS1-1)

1 Pour a spoonful of salt into the water. Stir until you can no longer see the salt. Record your observations.

2 Use the dropper to place three separate drops of salt water onto black paper.

3 Let the water evaporate for one hour or more.

4 Use the hand lens to examine the three areas of the paper where you dropped the salt water. Record your observations.

The smaller the crystals of salt are, the more quickly they dissolve in water.

Wrap It Up!

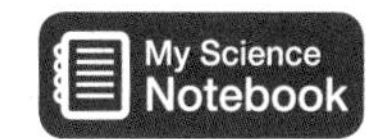

1. **Describe** What did you observe on the paper after the water evaporated?
2. **Explain** How does the size of the salt particles change when salt is dissolved in water?
3. **Explain** How do your results provide evidence that matter is made of particles too small to see?

Develop a Model

You have observed evidence that particles of matter exist even when they are not visible. When you blew up a balloon, particles of air pushed on the inside of the balloon and filled it up. When you let salt water evaporate, particles of salt were revealed. Now it's your turn. Imagine that you want to explain what you have learned to a friend. What model can you use to explain that matter is made of small particles too small to be seen?

1. **Construct an explanatory model.** 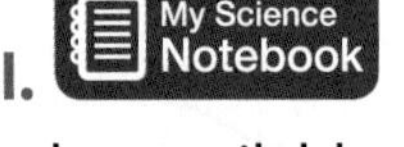

   Study the materials your teacher makes available to you. Think about how you could use these materials to show evidence that matter is made of particles too small to be seen. Draw and label a picture of your model. Explain how your model will work.

2. **Conduct an investigation.**
   After your teacher approves your design, build your model. Conduct an investigation that tests your model. Collect data from your test.

3. **Analyze results and revise your model.**
   Study your data. Does your model do what you want it to do? Revise your model and retest until you are satisfied with how it works.

4. **Share your model.**
   Use your model to explain to a partner that matter is made of particles too small to be seen. Does your partner "get" it? If not, work together to further improve your model. Then share your final model with the class.

  **PE** 5-PS1-1. Develop a model to describe that matter is made of particles too small to be seen.

Hot gas particles fill these balloons and allow them to float.

Stories in Science

Scientist and Role Model

Shirley Jackson worked to keep the use of nuclear energy safe. A nuclear reactor is shown to the right of a huge cooling tower.

Shirley's research led to development of fiber optic cables. These greatly improved communication technology.

In 2016, President Barack Obama awarded Shirley Jackson the National Medal of Science. That is the highest honor in the nation for contributions in science and engineering.

DCI PS1.A: Structure and Properties of Matter. Matter of any type can be subdivided into particles that are too small to see, but even then the matter still exists and can be detected by other means. A model shows that gases are made from matter particles that are too small to see ahd are moving freely around in space can explain many observations, including the inflation and shape of a balloon and the effects of air on larger particles or objects. (5-PS1-1)

NS Scientific Knowledge Assumes an Order and Consistency in Natural Systems. Science assumes consistent patterns in natural system. (5-PS1-2)

In 2005, *Time* magazine described Shirley Jackson as "... perhaps the ultimate role model for women in science."

Shirley Ann Jackson was inspired by watching the Russian satellite Sputnik being launched in space in 1957. She knew she wanted to study science. At the time, she was only 11 years old. Her parents supported her interest in science. But she was struggling to find a school in Washington, D.C., that would challenge her. Even after the 1954 Supreme Court ruling to desegregate schools, schools in Shirley's area were not nurturing her love of science. Despite the adversity she faced while obtaining her education, Shirley remained undeterred. She was the first African-American woman to graduate with a doctorate in theoretical physics from the Massachusetts Institute of Technology (MIT).

Not only was she one of few women in her class at MIT, Shirley was one of only two African-American women at the time. She often worked alone at school, but after graduating, her career took off. She applied her brilliant mind and knowledge to study some of the tiniest particles of matter. Her research helped develop technologies with which you are familiar. These include caller ID and call waiting on phones, solar cells, and fiber optic cables that transmit gigabytes of data all over the world in a matter of seconds.

Shirley was appointed as chairman of the U.S. Nuclear Regulatory Commission (NRC) by President Clinton in 1995. The NRC works to keep the use of nuclear energy safe. This includes reactor safety and the proper disposal of nuclear waste as well as making sure nuclear plants are secure. In 2017, there were 61 nuclear power plants operating in the United States. These plants provide about 20 percent of electricity used in the U.S. The appointment to chairman was another huge success for Shirley, as well as another first. She was the first African-American woman to hold the position.

Shirley continued her career of firsts. She became the first African-American woman to be the president of Rensselaer Polytechnic Institute (RPI). This university is known for its strength in technology education. Appointed in 1999, Shirley still holds her post as president of RPI. Shirley has also taught and served on President Obama's Intelligence Advisory Board. She has won numerous awards throughout her career. In 2005, *Time* magazine described Shirley Jackson as "... perhaps the ultimate role model for women in science." Indeed, Shirley is an inspiration for young girls who are interested in studying science, math, engineering, and technology. She did not let gender or race barriers get in the way of achieving her goals and living her dreams.

Wrap It Up!

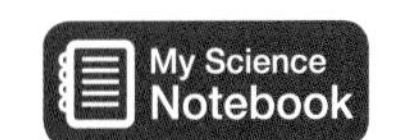

1. **Explain** What do you think is Shirley Jackson's most important acheivement? Tell why you think it is important.

2. **Apply** How has Shirley Jackson's research led to changes that affect you in everyday life?

Properties of Matter

Your friend asks you to close your eyes. In your hand, he places a large, polished rock. Even before you open your eyes, you know the object is a rock based on your sense of touch. The rock's large mass and hard, smooth surface give it away. Scientists can measure the properties of the rock more precisely to identify exactly what type of rock it is.

Physical **properties** are observable characteristics of a material that identify the material. Look at the different objects on these pages. Each has observable characteristics. Often those characteristics make the objects practical for a specific use.

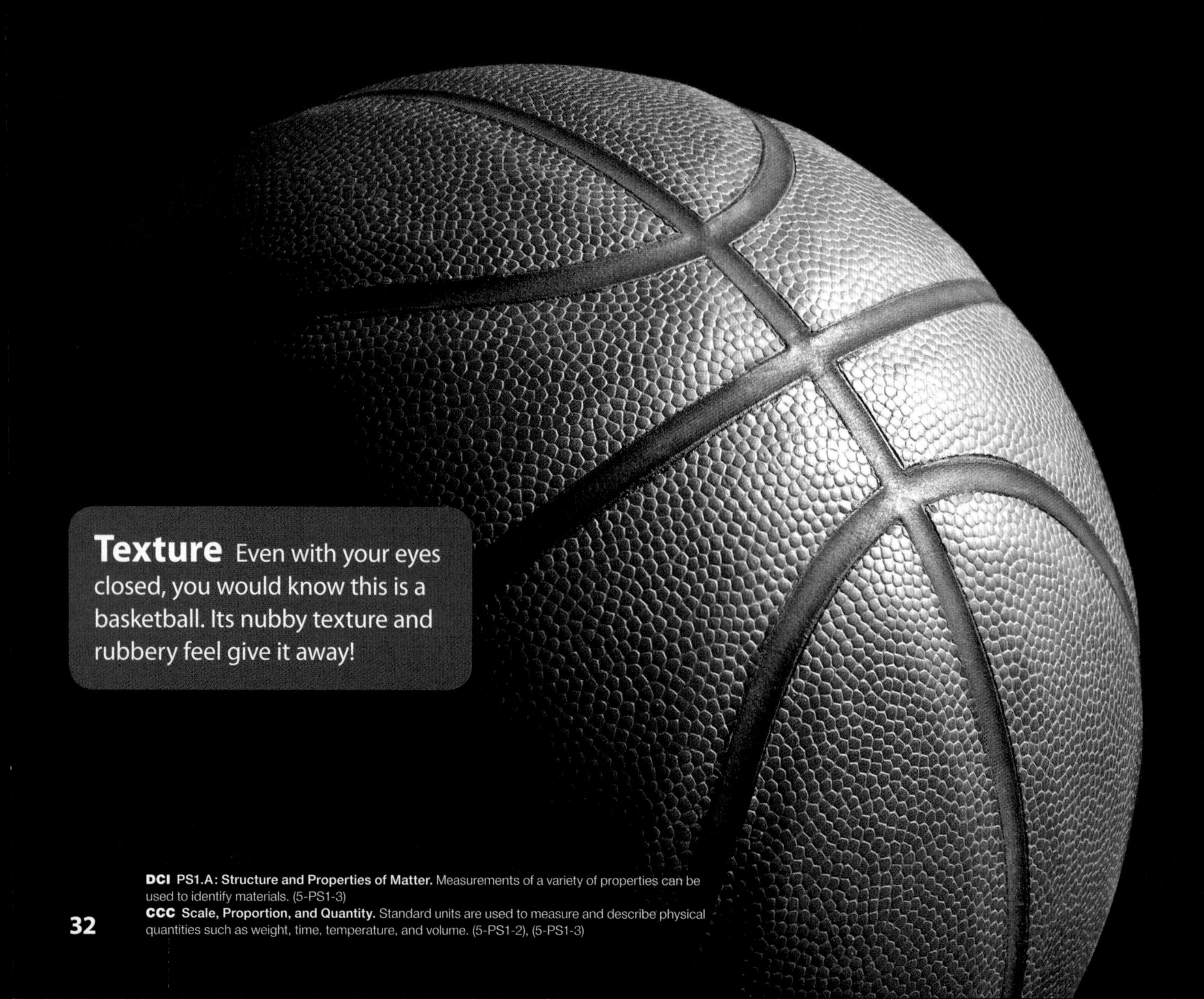

Texture Even with your eyes closed, you would know this is a basketball. Its nubby texture and rubbery feel give it away!

DCI PS1.A: Structure and Properties of Matter. Measurements of a variety of properties can be used to identify materials. (5-PS1-3)
CCC Scale, Proportion, and Quantity. Standard units are used to measure and describe physical quantities such as weight, time, temperature, and volume. (5-PS1-2), (5-PS1-3)

Color and Shape When you search for your purple helmet, you are using color to describe an object. Its round shape allows it to fit snugly on your head.

Hardness Could you use a stuffed toy to drive a nail into wood? Of course not! You need something hard and strong, such as this hammer.

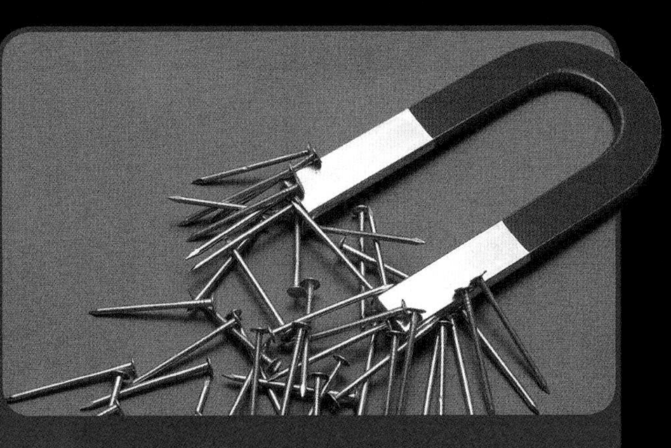

Magnetism The iron in these nails is attracted to the magnet. Magnets also attract cobalt and nickel.

Reflectivity What do mirrors and these shiny pots have in common? They all reflect light in a way that allows you to see an image.

Solubility The property of solubility allows you to mix up a cold glass of grape drink. The powder dissolves in the water.

Wrap It Up!

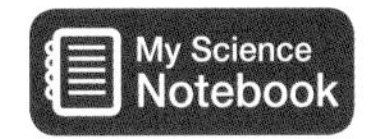

1. **List** Name six physical properties that can be used to identify matter.
2. **Apply** Choose an object in your surroundings. Describe its physical properties.
3. **Explain** Many properties can be measured. How could you measure the strength of different magnets?

Hardness

Hardness is a measure of how resistant a material is to scratching, bending, or denting. Hardness is measured on a scale that ranks materials from very soft to very hard. Scientists use hardness as a way to identify minerals. The chalk this artist is using to create the image on the concrete sidewalk contains a soft mineral. It rubs off on the sidewalk as the artist draws. The same would not be true if the artist tried to draw with a diamond!

The pavers are harder than the chalk. When scratched across the pavers, the softer chalk leaves a mark.

DCI PS1.A: Structure and Properties of Matter. Measurements of a variety of properties can be used to identify materials. (5-PS1-3)

CCC Scale, Proportion, and Quantity. Standard units are used to measure and describe physical quantities such as weight, time, temperature, and volume. (5-PS1-2), (5-PS1-3)

Diamond is the hardest mineral known. The diamond would actually scratch the concrete!

The chalk does not scratch the concrete.

SCIENCE in a SNAP

Test for Hardness

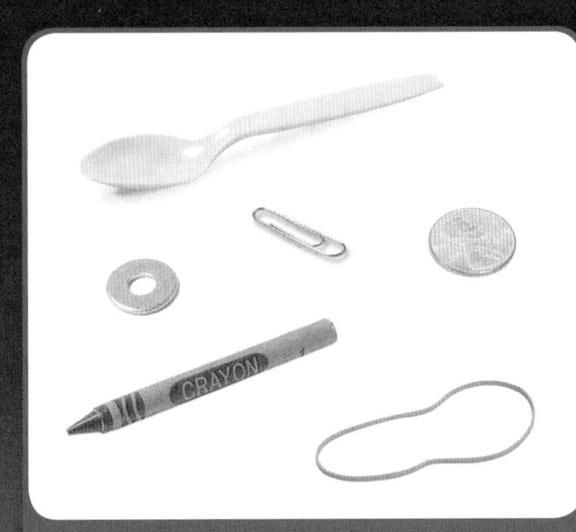

Scientists use the scratch test to determine the relative hardness of a material. If one material scratches another, it is harder than the other material.

Gather materials such as the ones shown here. Test each object to see if it can scratch the others. Put the objects in a row from softest to hardest.

Wrap It Up!

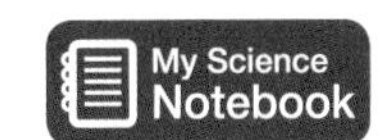

1. **Explain** How can the property of hardness be tested? Explain how the test works.
2. **Explain** How can you tell which is harder, chalk or concrete?
3. **Apply** Your fingernail is harder than chalk. Would chalk scratch your fingernail, or would your fingernail scratch chalk?

Investigate

Hardness

Pyrite looks a lot like gold, but pyrite is much harder than gold.

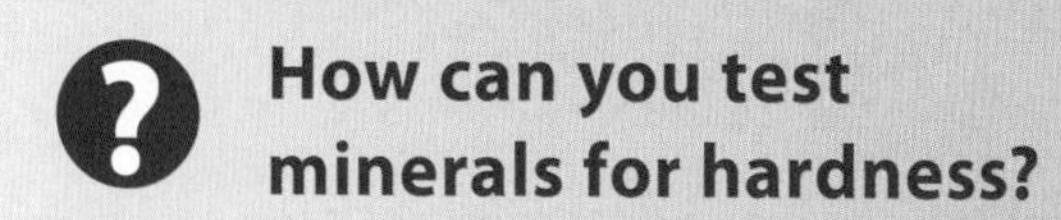

How can you test minerals for hardness?

You find a shiny golden rock in a stream. Could it be gold? Or is it pyrite, a material also known as fool's gold? The color and shininess of gold and pyrite are similar, but they differ in hardness. One quick way to determine the hardness of a material is to do a scratch test. In this investigation, you'll use the physical property of hardness to test some common minerals.

Materials

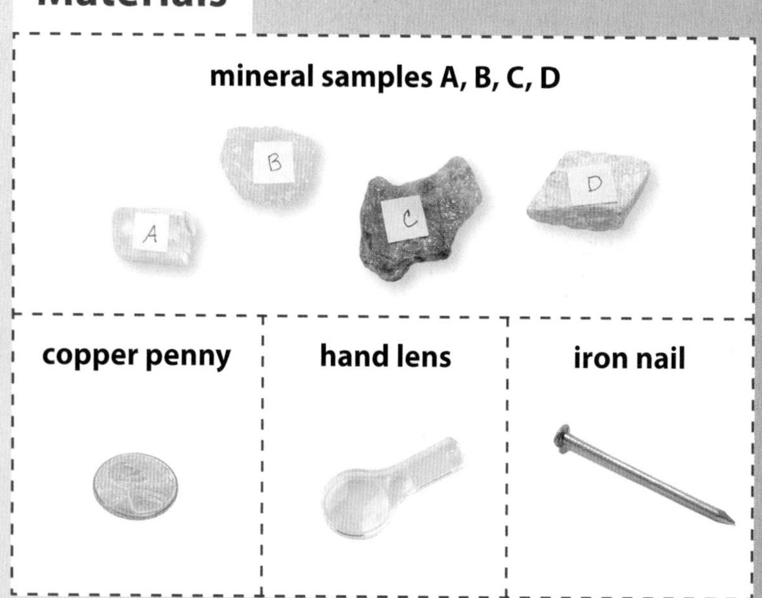

DCI PS1.A: Structure and Properties of Matter. Measurements of a variety of properties can be used to identify materials. (5-PS1-3)
CCC Scale, Proportion, and Quantity. Standard units are used to measure and describe physical quantities such as weight, time, temperature, and volume. (5-PS1-2), (5-PS1-3)

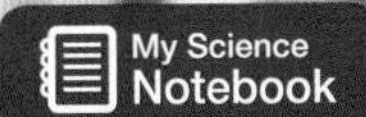

1 Examine each of the mineral samples with a hand lens. Record what you see.

2 Try to scratch each of the samples with your fingernail. Use the hand lens to examine the area you tried to scratch. Record your observations.

3 Repeat step 2, but this time scratch with the penny and then the iron nail. Record your observations.

4 Finally, try to scratch each sample with the other three samples. Record your observations.

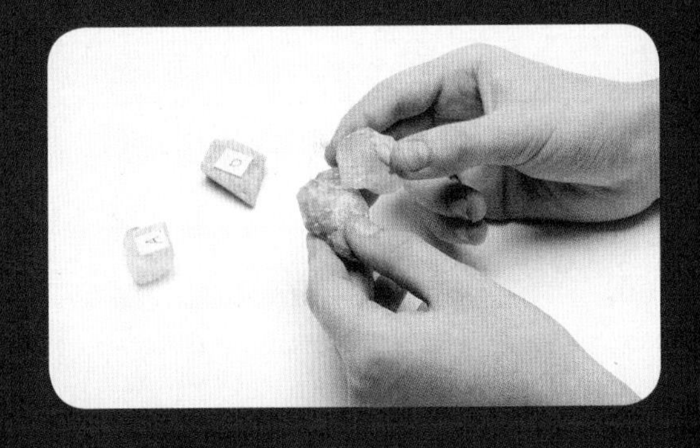

Wrap It Up!

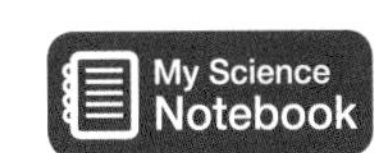

1. **Identify** Put your mineral samples in order from softest to hardest.
2. **Describe** What evidence did you use to order the samples?
3. **Apply** Gold is slightly harder than a fingernail. Pyrite is harder than a nail. Where would they fit in your ordered samples?

Magnetism

Another physical property of matter is magnetism. **Magnetism** is a force produced by magnets that pulls some metals. Objects can be identified based on whether they are attracted to a magnet.

If you have ever tried to pick up different objects with a magnet, you have seen that some materials are affected by magnetic force while others are not. Metals such as iron, nickel, and cobalt are magnetic. Most other materials are not magnetic.

DCI PS1.A: Structure and Properties of Matter. Measurements of a variety of properties can be used to identify materials. (5-PS1-3)
CCC Scale, Proportion, and Quantity. Standard units are used to measure and describe physical quantities such as weight, time, temperature, and volume. (5-PS1-2), (5-PS1-3)

SCIENCE in a SNAP

Magnetism

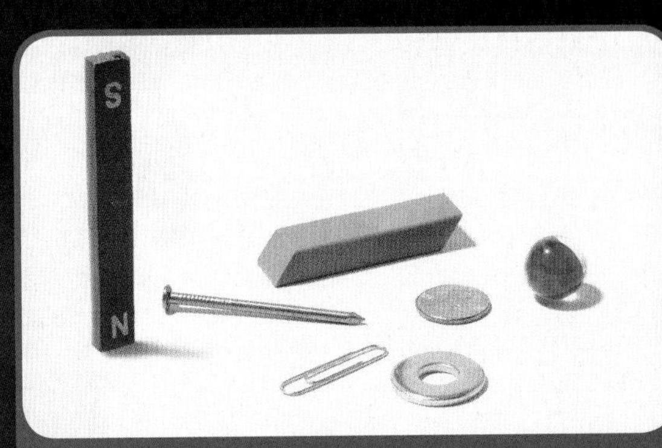

1 Observe a group of objects. Predict which objects will be attracted to a magnet. Sort the objects into two piles—ones you think are magnetic and ones you think are not magnetic.

2 Use the magnet to try to pick up each object. Record your observations.

? **Did your results support your predictions? Did any of your results surprise you?**

Maglev trains do not have wheels that roll on rails. Magnetism is used both to levitate, or lift, the train above the track and to move the train forward.

Wrap It Up!

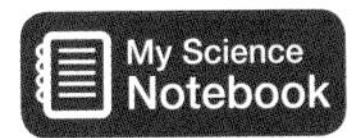

1. **Recall** What is magnetism?
2. **Identify** Give an example of an item that is magnetic. What type of metal does it most likely contain?
3. **Explain** How can the property of magnetism be tested?

Electrical Conductivity

Electrical conductivity is another property of matter. **Electrical conductivity** is a measure of how well electricity can move through a material. Good conductors of electrical energy, or **electrical conductors,** allow electricity to flow easily. Metals such as copper, gold, silver, and iron are good electrical conductors. Copper is commonly used to make electrical wires because it is such a good conductor.

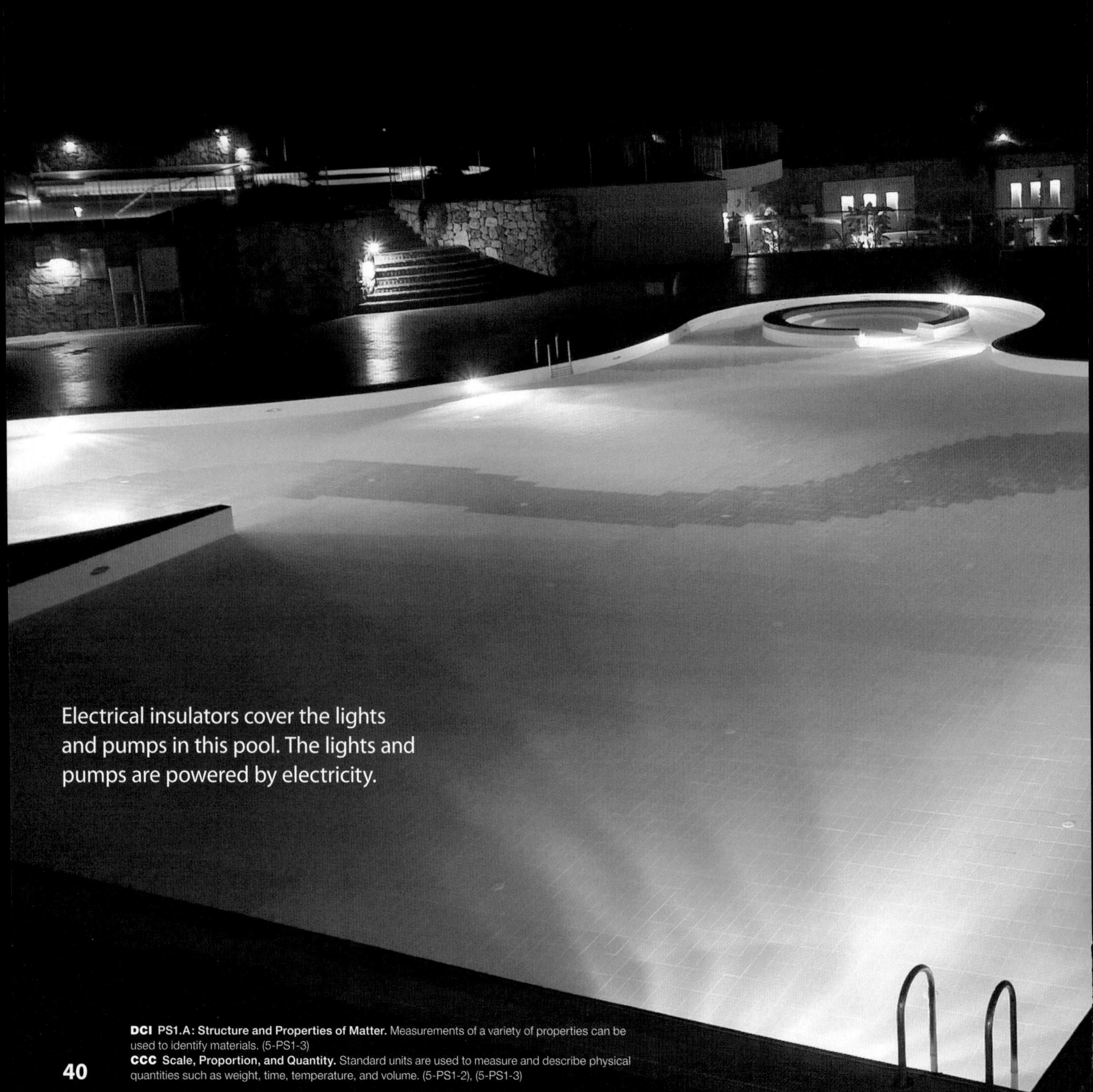

Electrical insulators cover the lights and pumps in this pool. The lights and pumps are powered by electricity.

DCI **PS1.A: Structure and Properties of Matter.** Measurements of a variety of properties can be used to identify materials. (5-PS1-3)
CCC **Scale, Proportion, and Quantity.** Standard units are used to measure and describe physical quantities such as weight, time, temperature, and volume. (5-PS1-2), (5-PS1-3)

Because electricity can be dangerous, it is important to protect people from it. An **electrical insulator** is a material that slows or stops the flow of electricity. Plastic, rubber, wood, and glass are good electrical insulators.

ELECTRICAL CONDUCTORS

Copper wires carry electricity in power lines and electrical plugs.

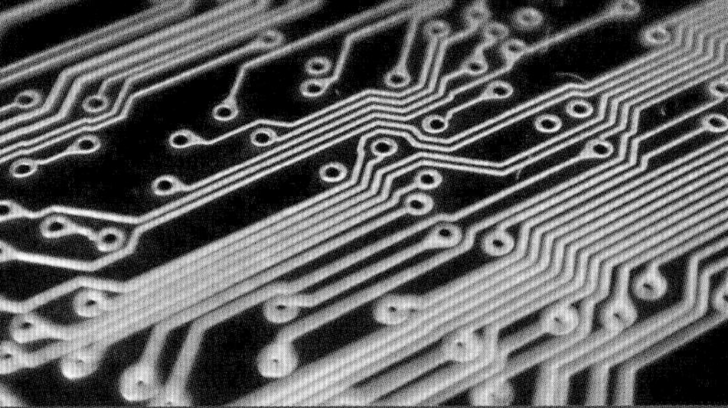

Gold carries electricity in some parts of a computer.

ELECTRICAL INSULATORS

Plastic coating on wires helps to prevent electrical shocks.

Glass electrical insulators prevent electricity from reaching a person working on power poles.

Wrap It Up!

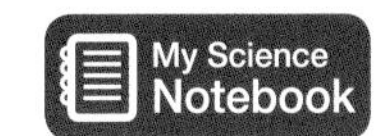

1. **Contrast** What is the difference between an electrical conductor and an electrical insulator?
2. **Recall** From what you've learned, what are two different properties of iron?
3. **Apply** Explain why electrical gloves are made of rubber.

Investigate

Electrical Conductivity

? Which materials conduct electricity?

You will never see electrical wires made of wood. Wood just isn't a good conductor. Wood is, however, a good insulator. How well a material conducts electricity is a physical property of the material. In this investigation, you'll test the ability of different materials to conduct electricity.

Materials

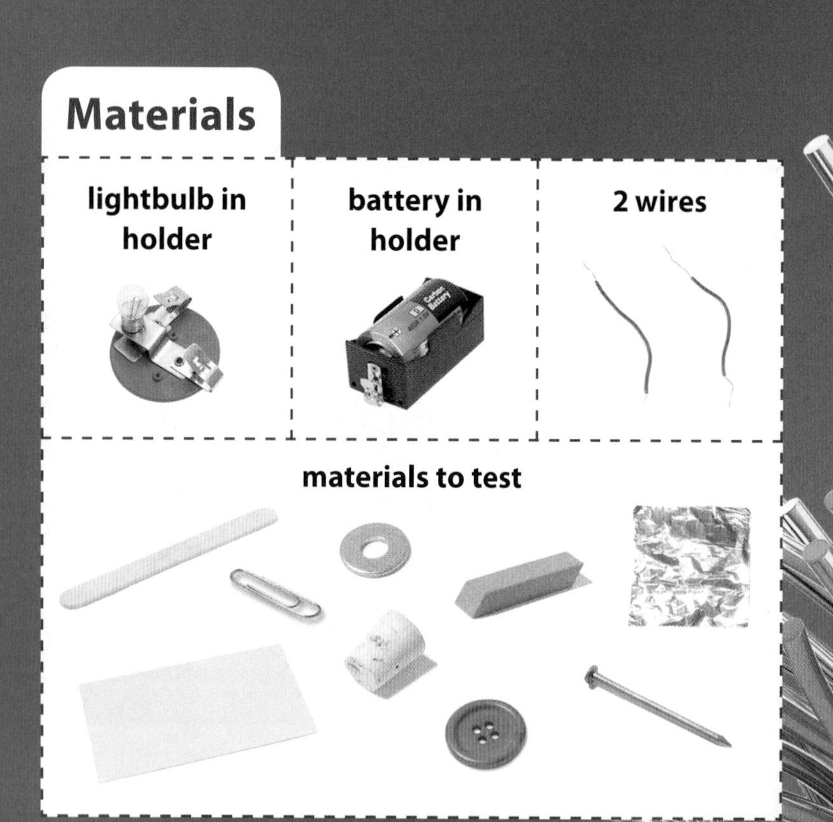

The plastic around the copper wire is a good insulator. The copper wire is a good conductor.

DCI PS1.A: Structure and Properties of Matter. Measurements of a variety of properties can be used to identify materials. (5-PS1-3)
CCC Scale, Proportion, and Quantity. Standard units are used to measure and describe physical quantities such as weight, time, temperature, and volume. (5-PS1-2), (5-PS1-3)

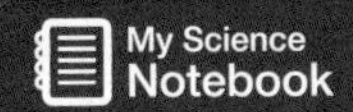

1 Examine the materials to test. Predict which materials will conduct electricity. Record your predictions in your science notebook.

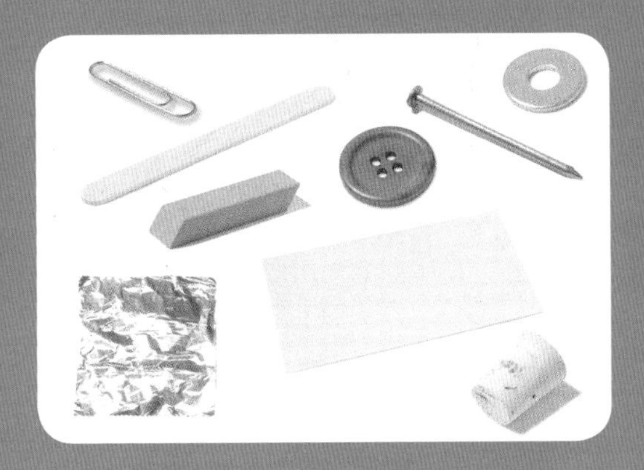

2 Attach the first wire to one end of the battery holder by connecting it to the metal piece. Attach the other end of the wire to the bulb holder. Attach the second wire to the other end of the battery holder.

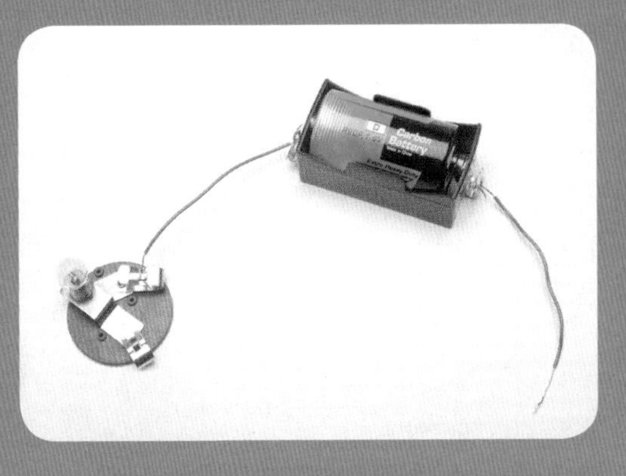

3 Wrap the end of the free wire around the nail. Touch the end of the nail to the open end of the bulb holder. Record your observations.

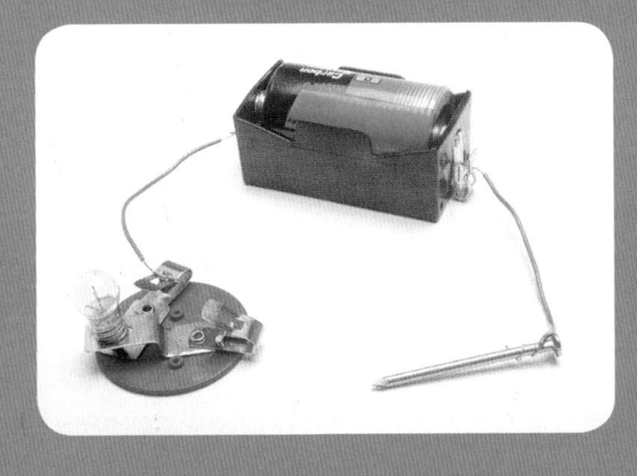

4 Use the free end of the wire and bulb holder to test the rest of the materials. Record your observations.

Wrap It Up!

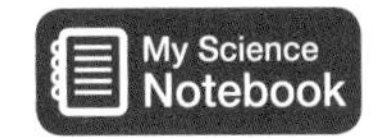

1. **Explain** Did your results support your predictions? Explain.
2. **Classify** Identify each material as an electrical conductor or an electrical insulator.
3. **Apply** What other materials can you think of that would be good electrical conductors and electrical insulators? How could you test them to see?

Thermal Conductivity

Matter is made up of particles that are always vibrating, or moving. The energy of moving particles is called **thermal energy.** The ability to conduct thermal energy is a physical property that can be used to identify materials.

Good conductors of thermal energy, or **thermal conductors,** allow thermal energy to flow easily through them as heat. Metals such as copper, aluminum, and iron are good thermal conductors.

The metal rod on the thermometer conducts thermal energy.

Iron is a good thermal conductor. Many pots and pans are made of metals, including iron.

This spatula is made of wood. Wood is a good thermal insulator.

A stove burner provides a heat source to cook food.

DCI PS1.A: Structure and Properties of Matter. Measurements of a variety of properties can be used to identify materials. (5-PS1-3)
CCC Scale, Proportion, and Quantity. Standard units are used to measure and describe physical quantities such as weight, time, temperature, and volume. (5-PS1-2), (5-PS1-3)

Cloth, wood, and rubber objects do not conduct thermal energy well. These materials are **thermal insulators.** We can use these materials to protect us from hot objects, such as a pot on a stove. Glass, plastic, and leather are also good thermal insulators.

Pot holders are made of cloth. Cloth is a good thermal insulator.

SCIENCE in a SNAP

Test Thermal Conductors

1 Predict which spoon will be a better thermal conductor. Record your predictions.

2 Test your prediction. Place the spoons in a cup of very warm water. Feel the stems of the spoons after a few minutes. Record your observations.

How did the thermal conductivity of the spoons differ?

Wrap It Up!

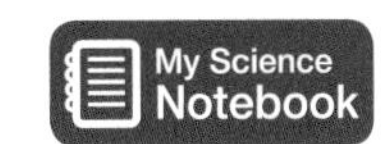

1. **Contrast** What is the difference between a thermal conductor and a thermal insulator?
2. **Classify** Identify the following materials as thermal conductors or thermal insulators: a wooden spoon, an iron frying pan, a plastic spatula, a steel fork.
3. **Apply** Explain why ceramic cups for drinks like coffee, tea, or cocoa usually have handles.

Investigate

Solubility

Which materials dissolve in water?

When sugar is stirred into lemonade, the sugar dissolves. The mixture of sugar and lemon juice is called a **solution.** In a solution, dissolved particles are distributed evenly and you can no longer see them. The ability of one substance to dissolve in another, or **solubility,** is a physical property of matter. In this investigation, you'll test the solubility of materials in water.

Materials

DCI PS1.A: Structure and Properties of Matter. Measurements of a variety of properties can be used to identify materials. (5-PS1-3)

CCC Scale, Proportion, and Quantity. Standard units are used to measure and describe physical quantities such as weight, time, temperature, and volume. (5-PS1-2), (5-PS1-3)

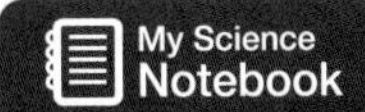

1 Predict what will happen when sand is added to water. Add a half spoonful of sand to a cup with water. Stir the water for about 30 seconds. Record your observations in your science notebook.

2 Predict what will happen when salt is added to water. Add a half spoonful of salt to a cup with water. Stir the water for about 30 seconds. Record your observations in your science notebook.

3 Predict whether lemon juice is soluble in water. Then predict whether vegetable oil is soluble in water. Record your predictions.

4 Pour 25 mL of lemon juice into the third cup of water. Stir for about 30 seconds. Repeat using vegetable oil and the fourth cup of water. Record your observations.

What do you think lemonade tastes like if you take a sip before the sugar dissolves?

Wrap It Up!

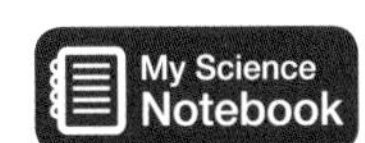

1. **Predict** Did your results support your predictions? Explain.
2. **Classify** Identify each material used in this investigation as soluble or insoluble in water.
3. **Apply** What other materials can you think of that might be soluble in water? How could you test to see?

Heating

Matter can change states when it is heated. For example, when ice is heated to its **melting point,** it melts into water. When water is heated to its **boiling point,** it becomes a gas called water vapor. Whether water is a solid, liquid, or gas, it is still water. These changes in state from a solid to a liquid and back again are known as **physical changes.** Physical changes, such as boiling water or cutting paper with a pair of scissors, do not change the material into a different one.

At its boiling point, water begins to change from a liquid to a gas. This change occurs at a temperature of 100°C (212°F).

DCI **PS1.A: Structure and Properties of Matter.** The amount (weight) of matter is conserved when it changes form, even in transitions in which it seems to vanish. (5-PS1-2)
CCC **Scale, Proportion, and Quantity.** Standard units are used to measure and describe physical quantities such as weight, time, temperature, and volume. (5-PS1-2), (5-PS1-3)

As water changes and rises from a boiling pot, it seems to vanish. But the water is still there, just in a different form. Even when matter changes state, the amount of matter stays the same. This principle is called the **conservation of matter**.

The melting point of water is 0°C (32°F). If you heat frozen water, it begins to change into a liquid a 0°C (32°F).

Wrap It Up!

1. **Recall** As you add heat, at what temperature does frozen water melt? At what temperature does liquid water boil?
2. **Identify** Boiling and melting points are properties of matter. What are the boiling and melting points of water?
3. **Cause and Effect** How does boiling affect water's state of matter?

Cooling

Just as heating causes matter to change states, so does cooling. **Condensation** is the change from a gas to a liquid. As the water vapor over a pot of boiling water cools in the air, the water begins to condense. That produces the fog-like steam you see over the pot. Water vapor itself is invisible. But steam is actually tiny drops of liquid water! When the water vapor changes to steam, the amount of matter is conserved.

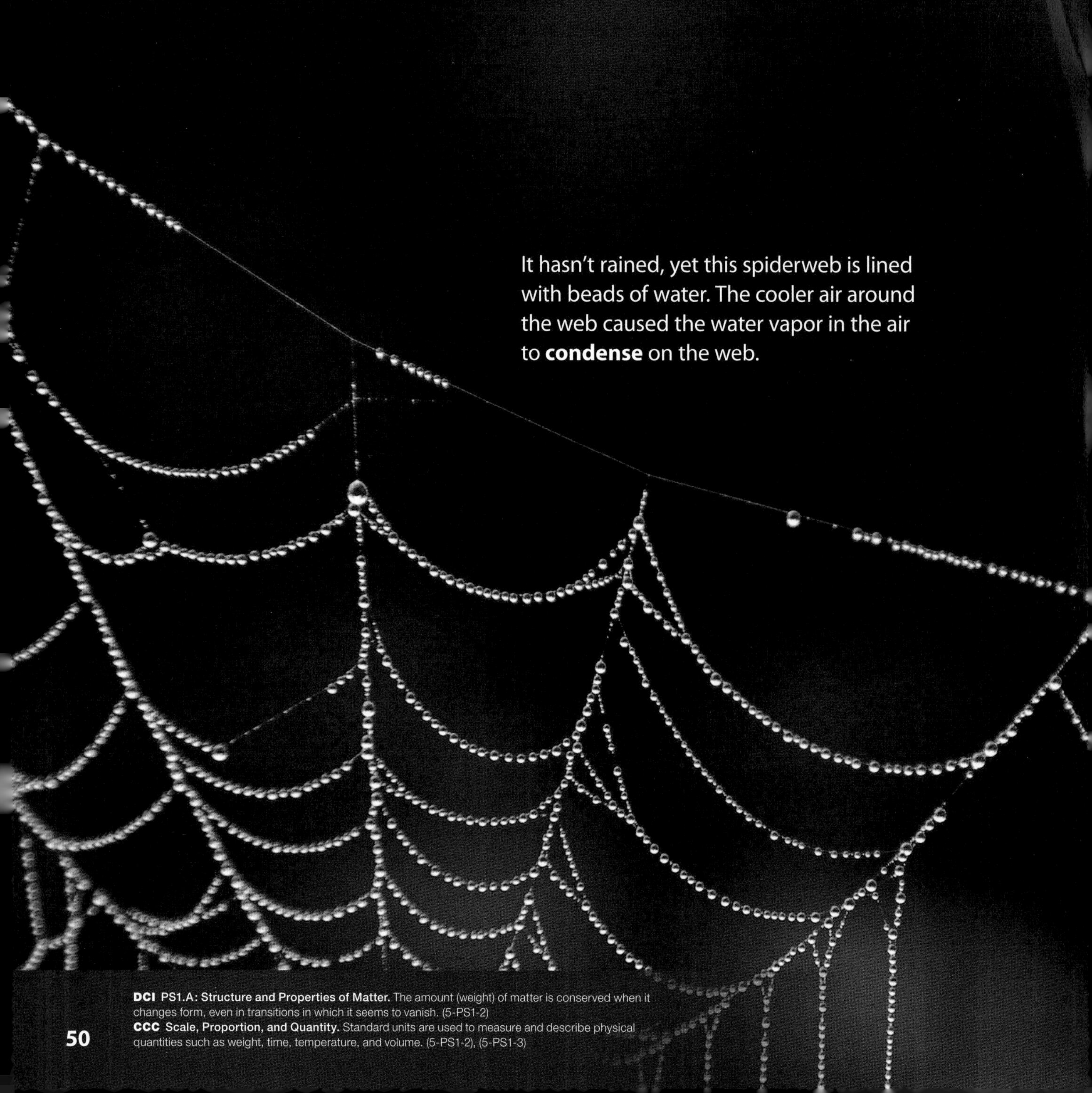

It hasn't rained, yet this spiderweb is lined with beads of water. The cooler air around the web caused the water vapor in the air to **condense** on the web.

DCI PS1.A: Structure and Properties of Matter. The amount (weight) of matter is conserved when it changes form, even in transitions in which it seems to vanish. (5-PS1-2)
CCC Scale, Proportion, and Quantity. Standard units are used to measure and describe physical quantities such as weight, time, temperature, and volume. (5-PS1-2), (5-PS1-3)

Water condenses when moist, warm air comes in contact with a cooler surface. That's why the bathroom mirror fogs up when you take a shower.

As liquid water is cooled to 0°C (32°F), it begins to freeze. Sometimes the air cools so quickly that water vapor from the air condenses and quickly freezes. Little bits of ice called frost form. Frost is common in areas that have a lot of moisture in the air and often occurs at night, when temperatures drop. When water vapor condenses and freezes, the amount of matter is conserved. Frost can form on the insides of windows in the winter, when the water vapor in the warm inside air comes in contact with a cold window.

The water on the outside of this cold glass didn't escape from the inside of the glass. It condensed from the water vapor in the surrounding air.

The frost on this glass is really water from the air that condensed and then froze.

Wrap It Up! 

1. **Define** What is condensation?
2. **Explain** How can a window that is not wet become covered with frost?
3. **Summarize** Make a diagram to show water's three states of matter and its change from one to another. Include the labels: *ice, water, water vapor, condensation, melting, freezing*.

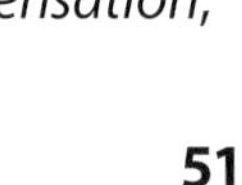

SCIENCE
TECHNOLOGY
ENGINEERING
MATH

Design a Lunch Box

It's bitterly cold, and the wind is blowing hard. But inside their quilted coats, the hikers are nice and warm. When the hikers stop for lunch, the cocoa in their thermos is steaming hot. What keeps the hikers and their cocoa so warm? Insulation.

As you have learned, thermal insulators are materials that do not conduct heat well. Cloth and glass are good thermal insulators. Air is an excellent insulator, especially when it's trapped between layers of cloth or inside small spaces.

Because insulation slows the flow of heat, it can serve two purposes. On a winter day, it can keep the thermal energy of warm objects from escaping into the cold air. But on a summer day, insulation can slow the flow of heat from the air into objects that people want to keep cool. In this project, you will use what you've learned about thermal insulators to design a lunch box that can keep food warm or cold.

PE 3–5-ETS1-2. Generate and compare multiple possible solutions to a problem based on how well each is likely to meet the criteria and constraints of the problem.
PE 3–5-ETS1-3. Plan and carry out fair tests in which variables are controlled and failure points are considered to identify aspects of a model or prototype that can be improved.
PE 5-PS1-3. Make observations and measurements to identify materials based on their properties.

The hiker's coat and her thermos are composed of materials that slow the flow of heat into the air.

The Challenge

Your engineering challenge is to design and build an insulated lunch box. Your lunch box must:

- keep one bottle of water cold
- keep a second bottle of water warm
- hold both bottles firmly in place

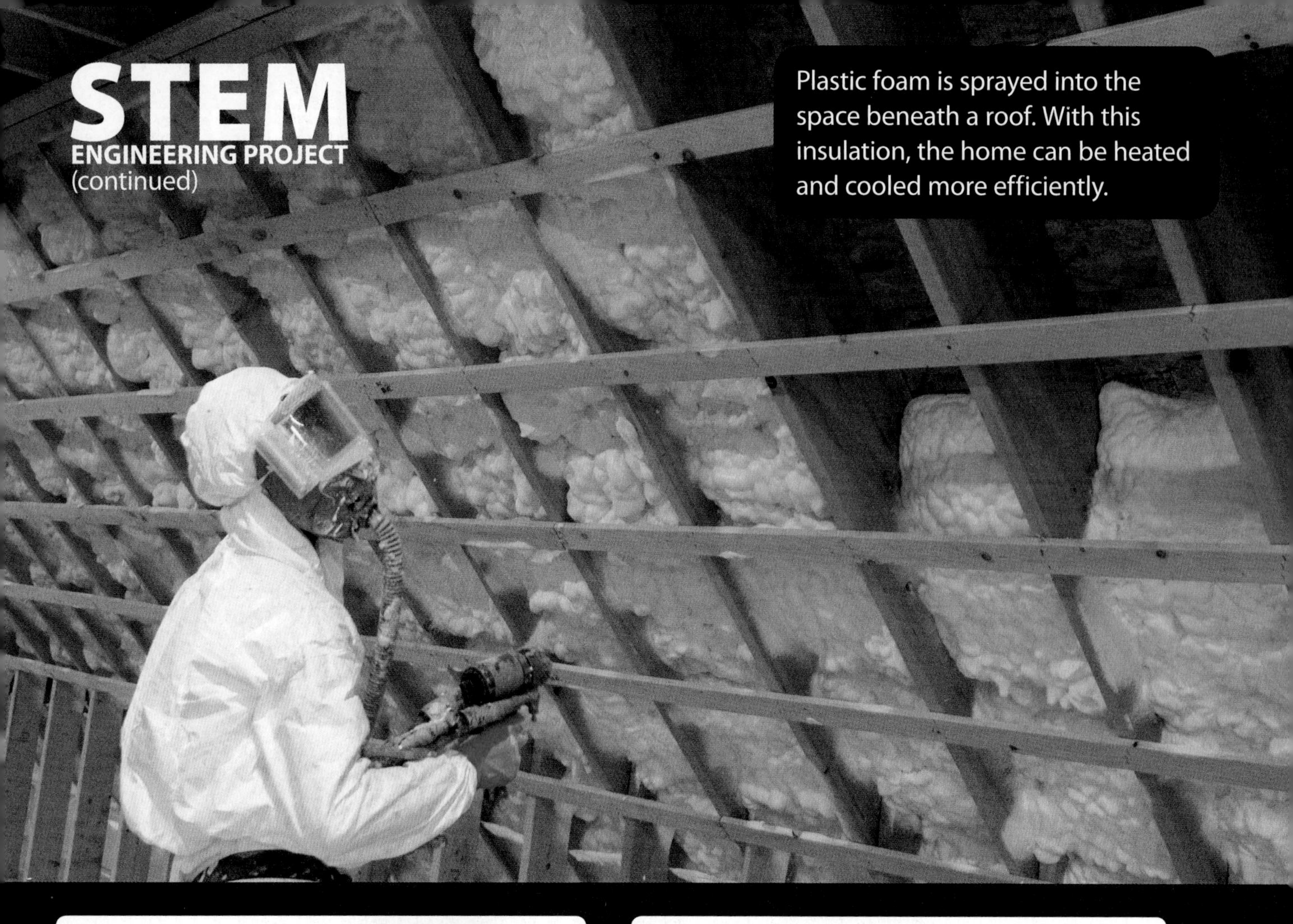

STEM
ENGINEERING PROJECT
(continued)

Plastic foam is sprayed into the space beneath a roof. With this insulation, the home can be heated and cooled more efficiently.

1 Define the problem.

Engineers start their work by defining the problem they need to solve. Think about your challenge. What does your lunch box need to do? You will know that your design is successful if your lunch box meets these criteria.

Engineers must also consider the constraints, or limits to the design. Constraints often involve the amount of space or the kinds of materials that are available. Your teacher will show you the materials you can use to build your lunch box. You cannot use any other materials. When you make your lunch box, all of the insulation and two bottles of water must fit inside a shoebox.

Write the problem you need to solve in your science notebook. Also list the criteria and constraints of the problem.

2 Find a solution.

Observe the materials you can use to build your lunch box. Select three materials that you think will be good insulators. Plan an investigation to determine which of them will keep water cold for the longest time. Your investigation will use three bottles of cold water.

- How will you insulate the bottles?
- What measurements will you make?
- How will you make your tests fair?

Write the steps of your investigation in your science notebook.

Conduct your tests, and record your data in a table. Make a line graph of your results.

Use the results from your tests to design your lunch box. Have each team member sketch a plan. Then select the plan that best meets the criteria and constraints of the problem. Draw a diagram of your design.

The foam insulation around this pipe has tiny pockets of air that slow the flow of heat from the pipe.

3 Test your solution.

Follow your plan to build your lunch box. Does all of the insulation fit inside? Does your lunch box hold two bottles firmly in place? As you build, make notes about any adjustments you make to your design.

With the class, plan a way to compare the lunch boxes built by all of the groups. When you have a plan, your teacher will give your group a bottle of hot water and a bottle of cold water. Test your design, and record your results.

Present your lunch box to the class. Share the results of your tests. Compare your results with those of the other teams. Which team's design was most effective in meeting the criteria and constraints of the problem? Discuss what is good and bad about each team's design.

4 Refine your solution.

Think about the results of your test. Did your lunch box meet the criteria and constraints of the problem? Also think about the other teams' designs. What materials did they use to insulate the hot and cold water? How could you use their ideas to improve your design?

Use what you have learned to make changes to your lunch box. Draw a diagram of your new design. In your science notebook, describe each change you made and explain why you made it.

Present your refined lunch box to the class. Compare your new design to your first design. Use evidence from your tests to explain why you made the changes.

If time allows, test your new design. Compare your results to those from the first test. Did your changes improve your design?

Investigate

Changing States of Water

How is matter conserved when water changes state?

You're already familiar with water in all three of its states: solid, liquid, and gas. Even when water changes from one form to another, its mass is conserved, or stays the same. However, some physical properties of water *do* change when it changes state. In this investigation, you'll observe some of these changes.

Materials

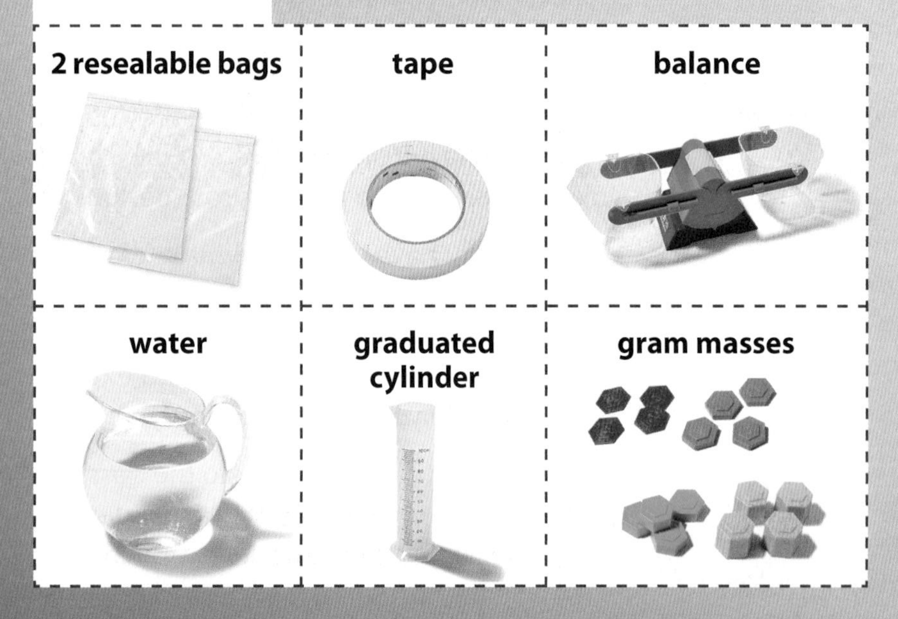

DCI PS1.A: Structure and Properties of Matter. The amount (weight) of matter is conserved when it changes form, even in transitions in which it seems to vanish. (5-PS1-2)
CCC Scale, Proportion, and Quantity. Standard units are used to measure and describe physical quantities such as weight, time, temperature, and volume. (5-PS1-2), (5-PS1-3)

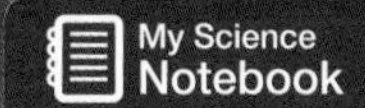

1 Label 2 plastic bags *Bag 1* and *Bag 2*. Use a graduated cylinder to measure 100 mL of water. Pour the water into the bag. Seal the bag. Repeat with the other bag.

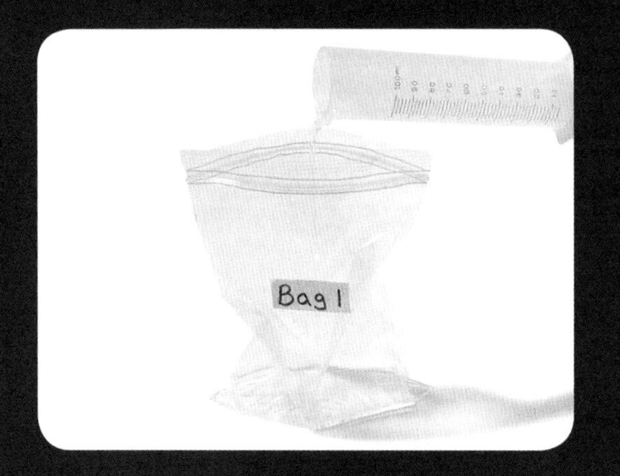

2 Use the balance and gram masses to measure the mass of each bag. Record your observations in your science notebook.

3 Predict what will happen when you freeze the water. Record your predictions. Place the bags in a freezer. The next day, take the bags out of the freezer. Measure the mass of the bags. Turn the bags in different directions and measure the mass again. Record your observations.

4 Place the bags in sunlight. Open Bag 1 only. Predict what will happen to the water in the bags after three days. Observe the bags every day for three days. Measure the mass of the bags after three days. Record your observations.

Wrap It Up!

1. **Predict** Did your results support your predictions? Explain.
2. **Compare and Contrast** Which properties of water stayed the same after cooling? Which properties changed?
3. **Infer** Explain the differences in the bags after step 4.
4. **Draw Conclusions** How do your findings demonstrate the conservation of matter?

Investigate

Mixtures

How is matter conserved when baking soda and water mix?

You have found that the amount of matter stays the same, even when it changes from one state to another. What happens when one material is mixed with another? Does the amount of matter stay the same after mixing? In science, a combination of materials is called a **mixture** when the materials do not change into something else after they are mixed. In this investigation, you'll determine whether matter is conserved when baking soda and water form a mixture.

Materials

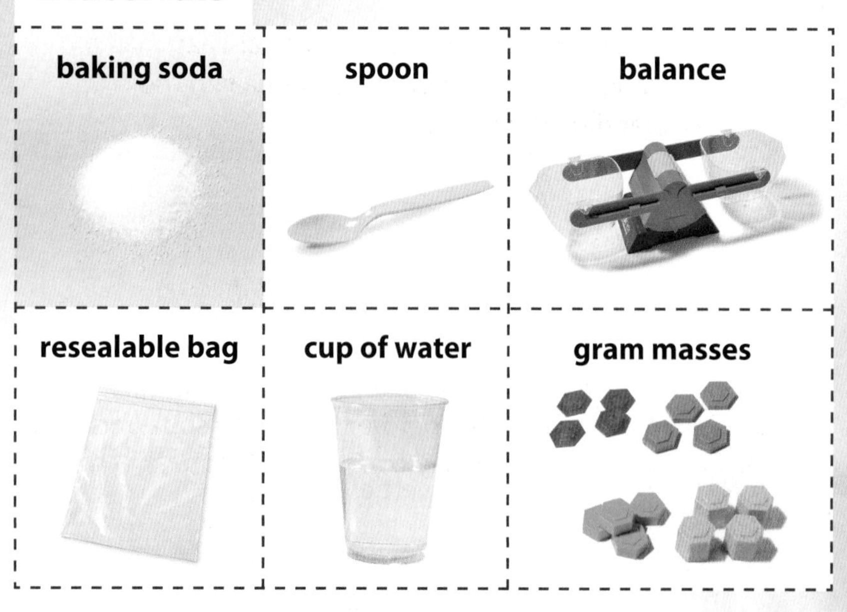

DCI PS1.A: Structure and Properties of Matter. The amount (weight) of matter is conserved when it changes form, even in transitions in which it seems to vanish. (5-PS1-2)
CCC Scale, Proportion, and Quantity. Standard units are used to measure and describe physical quantities such as weight, time, temperature, and volume. (5-PS1-2), (5-PS1-3)

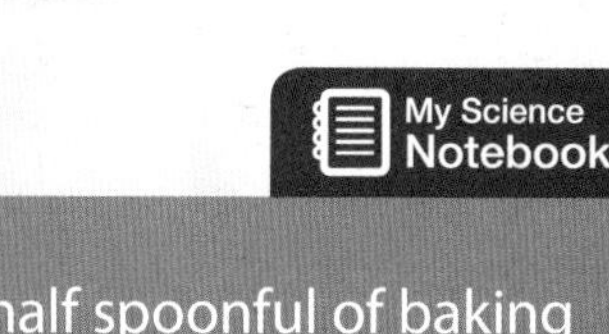

1 Measure a half spoonful of baking soda and put it into the bag. Seal the bag.

2 Place the bag of baking soda and the cup of water on the balance. Record the total mass of the materials.

3 Open the bag. Carefully empty the contents into the cup of water. Use the spoon to stir the mixture until the baking soda is completely dissolved in the water.

4 Place the cup of baking soda water and the empty bag on the balance. Record the total mass of the materials.

A fruit smoothie is a tasty mixture! The same amount of ingredients remain in the blender after they are mixed.

Wrap It Up!

1. **Compare** How did the mass of the materials before the baking soda and water were mixed compare to the mass of the materials after they were mixed?
2. **Analyze** Why was the empty bag added to the balance in step 4?
3. **Draw Conclusions** How do your findings demonstrate the conservation of matter?

Provide Evidence

You've observed that matter is conserved, even after it is changed by cooling, heating, or mixing. Now it's your turn to provide evidence of the conservation of matter. You'll develop an investigation to measure and graph the mass of matter before and after a physical change of your choice. Examples of changes could include heating, cooling, or mixing.

1. **Ask a question.** My Science Notebook
   How can you measure and graph quantities to provide evidence that matter is conserved?

2. **Plan and conduct an investigation.**
   Study the materials your teacher gives you. Think about how you can use some of the materials to provide evidence that the mass of matter stays the same after it undergoes a physical change. How will you measure the total mass of your material before and after this physical change? What type of physical change will you use? Record the steps of your plan in your science notebook. After your teacher approves your plan, carry out your investigation. Measure and record your data in a simple table.

3. **Analyze and interpret data.**
   Make a bar graph of your data. How do the quantities before and after the change compare? How can you explain your results? How do your findings provide evidence of the conservation of matter?

4. **Share your results.**
   Share your conclusions with the class. After hearing from your classmates, make a generalization about the conservation of matter.

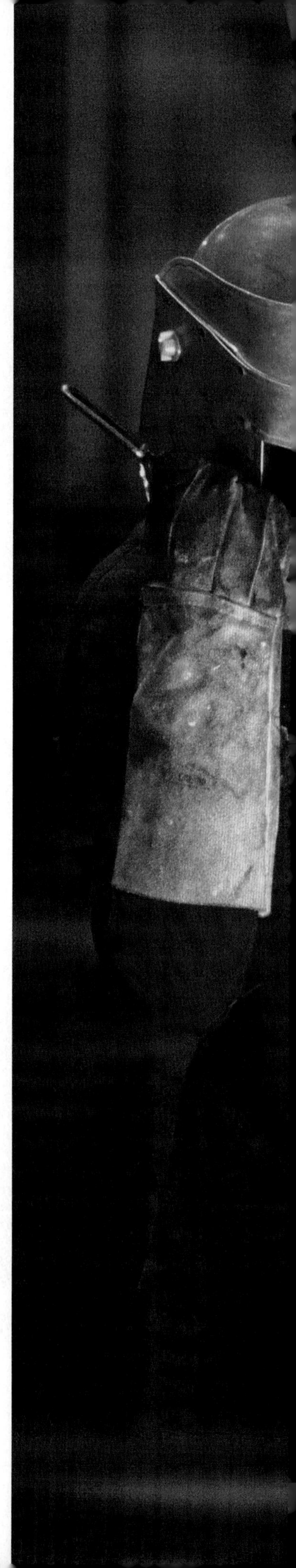

PE 5-PS1-2. Measure and graph quantities to provide evidence that regardless of the type of change that occurs when heating, cooling, or mixing substances, the total weight of matter is conserved.

This glowing liquid will harden into metal. The mass of the metal will remain the same.

Chemical Changes

Unlike the physical changes you have learned about so far, a **chemical change** causes a material to change into an entirely different material with properties that are different from the original material. The process by which a chemical change occurs is called a **chemical reaction.** In many chemical reactions, the new material cannot be changed back to the original material.

The conservation of matter applies to chemical changes as well as physical changes. The total mass of matter before a chemical reaction is the same as the total mass of matter after the reaction.

Inside a glow stick, a chemical reaction releases energy in the form of light.

DCI PS1.B: Chemical Reactions. When two or more different substances are mixed, a new substance with different properties may be formed. (5-PS1-4) • No matter what reaction or change in properties occurs, the total weight of the substances does not change. (5-PS1-2)
CCC Scale, Proportion, and Quantity. Standard units are used to measure and describe physical quantities such as weight, time, temperature, and volume. (5-PS1-2), (5-PS1-3)

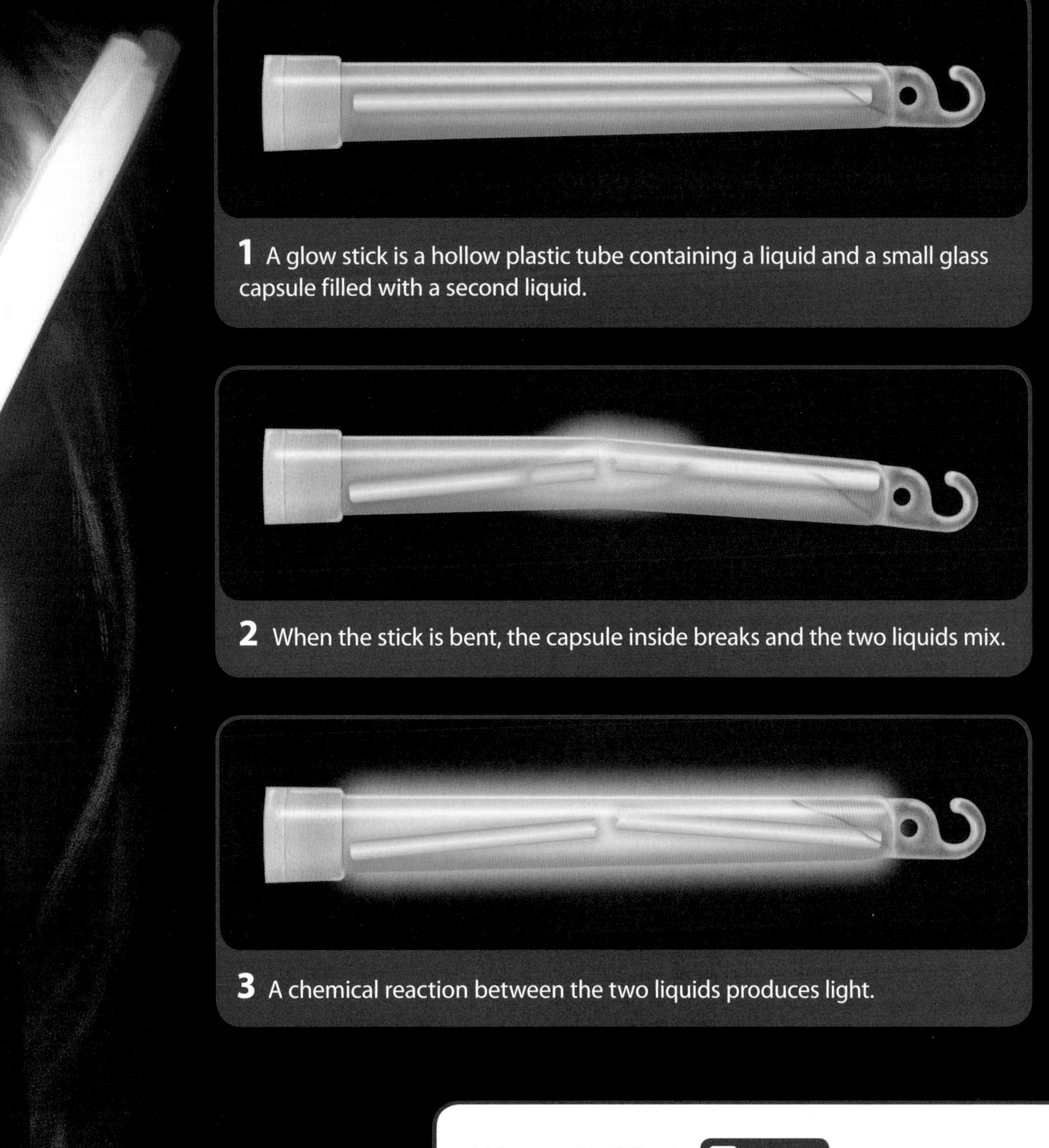

1 A glow stick is a hollow plastic tube containing a liquid and a small glass capsule filled with a second liquid.

2 When the stick is bent, the capsule inside breaks and the two liquids mix.

3 A chemical reaction between the two liquids produces light.

Wrap It Up!

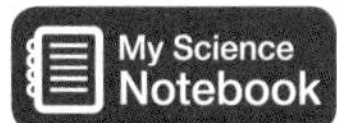

1. **Restate** What is a chemical reaction?
2. **Compare and Contrast** How is a chemical change alike and different from a physical change?
3. **Explain** What happens to the total mass of matter after a glow stick is bent and a chemical reaction takes place?

Signs of a Chemical Change

Chemical reactions are happening all around you all the time. For instance, chemical reactions digest food in your stomach. A banana that turns brown over time has undergone a chemical reaction. A burning candle is another example of a chemical reaction.

How do you know whether a chemical change has occurred? Learn to spot the signs. Some signs of chemical reactions are bubbles, changes in color, and production of smells, light, and heat.

The smell of burning toast and its dark black color tell you a chemical change has occurred. No matter what reaction or chemical change occurs, the total mass of the substances does not change.

DCI PS1.B: Chemical Reactions. When two or more different substances are mixed, a new substance with different properties may be formed. (5-PS1-4) • No matter what reaction or change in properties occurs, the total weight of the substances does not change. (5-PS1-2)
CCC Scale, Proportion, and Quantity. Standard units are used to measure and describe physical quantities such as weight, time, temperature, and volume. (5-PS1-2), (5-PS1-3)

The smell, the light, and the heat are three signs of chemical change in a burning sparkler.

When a weak acid is dropped on limestone, bubbles are produced. Bubbles are one sign of a chemical change.

Wrap It Up! My Science Notebook

1. **Identify** List five signs that a chemical reaction has occurred.
2. **Identify** What are some signs of a chemical reaction in a burning candle?
3. **Apply** Identify some chemical reactions that you observe regularly at home. What can you state about the total mass of the substances before and after each reaction?

Investigate

Chemical Reactions

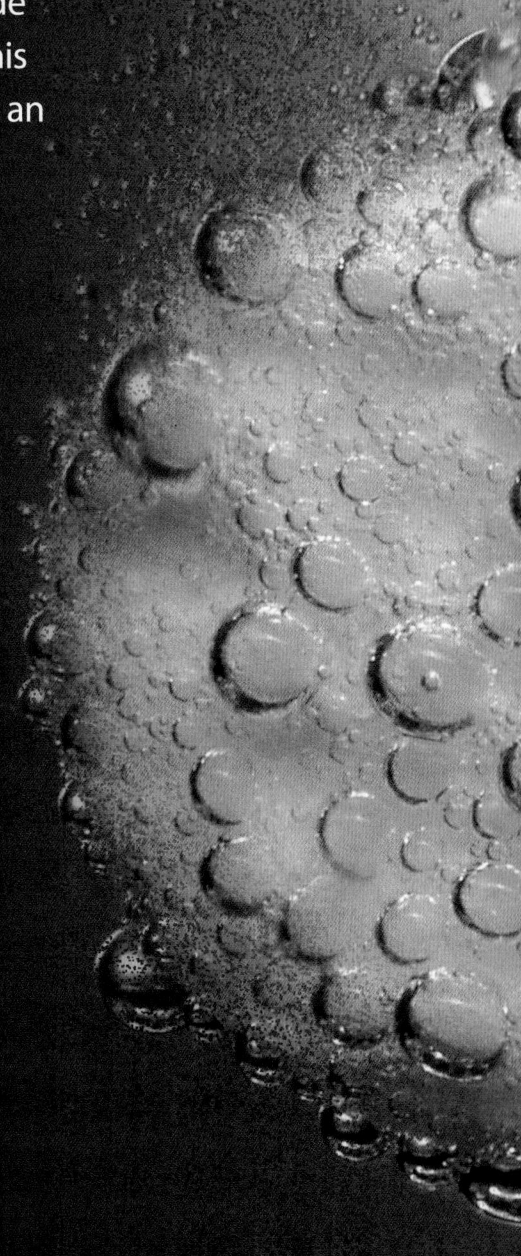

? How can you show that a new substance forms when some materials are mixed?

You can look for signs that chemical reactions are happening. The light and heat of a campfire, for instance, are signs that a chemical change is taking place. You can infer that the wood is changing into a different material. The wood is combining with oxygen from the air and changing into carbon dioxide gas, water, and other materials in the ash and smoke. In this investigation, you'll observe a chemical reaction between an effervescent tablet and water.

Materials

graduated cylinder	large resealable bag	balance

water	**effervescent tablet**	**gram masses**
		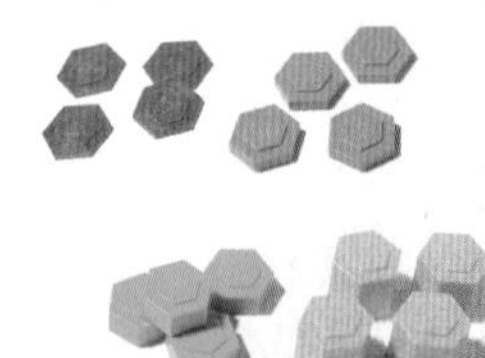

PE 5-PS1-4. Conduct an investigation to determine whether the mixing of two or more substances results in new substances.

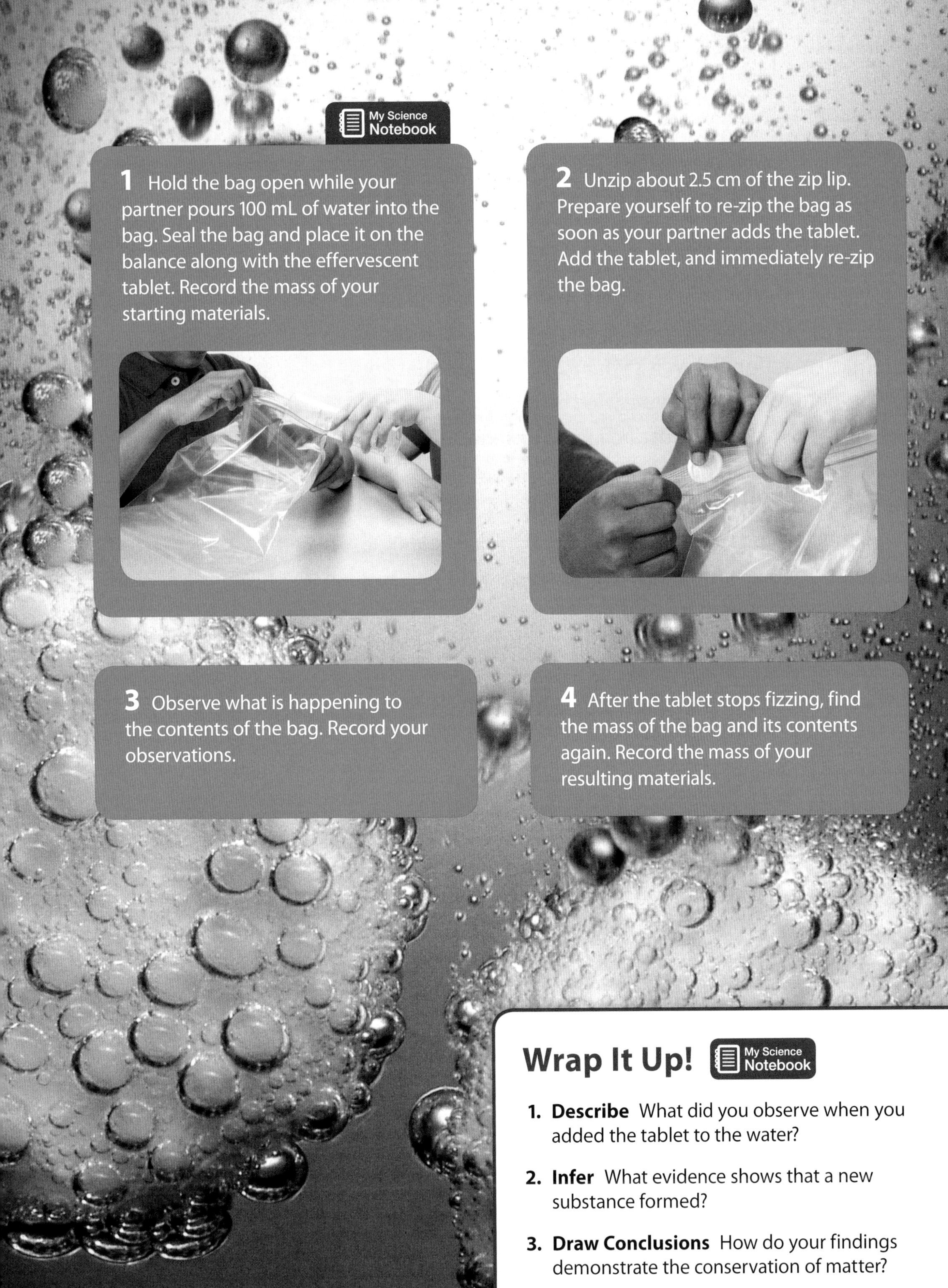

My Science Notebook

1 Hold the bag open while your partner pours 100 mL of water into the bag. Seal the bag and place it on the balance along with the effervescent tablet. Record the mass of your starting materials.

2 Unzip about 2.5 cm of the zip lip. Prepare yourself to re-zip the bag as soon as your partner adds the tablet. Add the tablet, and immediately re-zip the bag.

3 Observe what is happening to the contents of the bag. Record your observations.

4 After the tablet stops fizzing, find the mass of the bag and its contents again. Record the mass of your resulting materials.

Wrap It Up!

My Science Notebook

1. **Describe** What did you observe when you added the tablet to the water?
2. **Infer** What evidence shows that a new substance formed?
3. **Draw Conclusions** How do your findings demonstrate the conservation of matter?

Identify Materials

You've seen that materials can be identified by their properties. You've also seen how materials change as a result of heating and cooling, and what happens in chemical reactions. Now it's your turn. Imagine that you are helping to prepare dinner. To "help," someone has measured out sugar, cornstarch, baking soda, and baking powder. Unfortunately, the powders are not labeled! But these powders have different properties when mixed with liquids such as water, vinegar, and iodine. It will be your job to use the properties of these materials to identify which is which.

1. **Ask a question.** 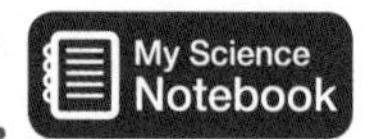

   How can you use observations and measurements to identify materials based on their properties?

2. **Plan and conduct an investigation.**
   Your teacher will provide four powder samples and three liquids. Study the powders and liquids. You should make careful observations, but do not taste any of the materials. Think about how you could test the powders to determine what they are. How will you mix each powder with each liquid? How will you measure the materials? How will you record your observations? Record the steps of your plan in your science notebook. After your teacher approves your plan, carry out your investigation. As you work, record your data in a chart. Include details on the characteristics of each powder and how each powder reacts to each of the liquids.

 PE 5-PS1-3. Make observations and measurements to identify materials based on their properties.

3. **Analyze your results.**
   After you conduct your tests, compare your results to the powder ID chart your teacher will provide. Use the chart and your data to determine which powder is which. Discuss the scientific basis of your findings. What evidence helped you to identify each powder?

4. **Share your results.**
   Share your conclusions with the class. As a class, discuss the observations that were most useful in identifying each powder.

Knowing how materials react enables firework makers to choose the right chemicals to burn, explode in certain ways, and produce colors as planned.

Research Scientist

Genghis Khan was a legendary ruler and warrior. In the early 13th century, he founded the largest empire in the known world. It is thought that he died in 1227 during a military battle, but the details of his death remain a mystery. His tomb has never been found.

That is where Dr. Albert Yu-Min Lin comes in. As a research scientist, he plans and carries out scientific investigations. Albert wants to know what happened to Genghis Khan, and he wants the public to help him. In his work with the Valley of the Khans Project, he uses a technique called crowdsourcing. This means he invites people all over the world to help him search through huge amounts of satellite imagery data for clues.

Albert uses a variety of different technologies in his hunt for the lost tomb. They include satellite imagery, ground-penetrating radar, and even electromagnets. He hopes to one day locate Genghis Khan's burial site and answer the many questions surrounding the ruler's mysterious death.

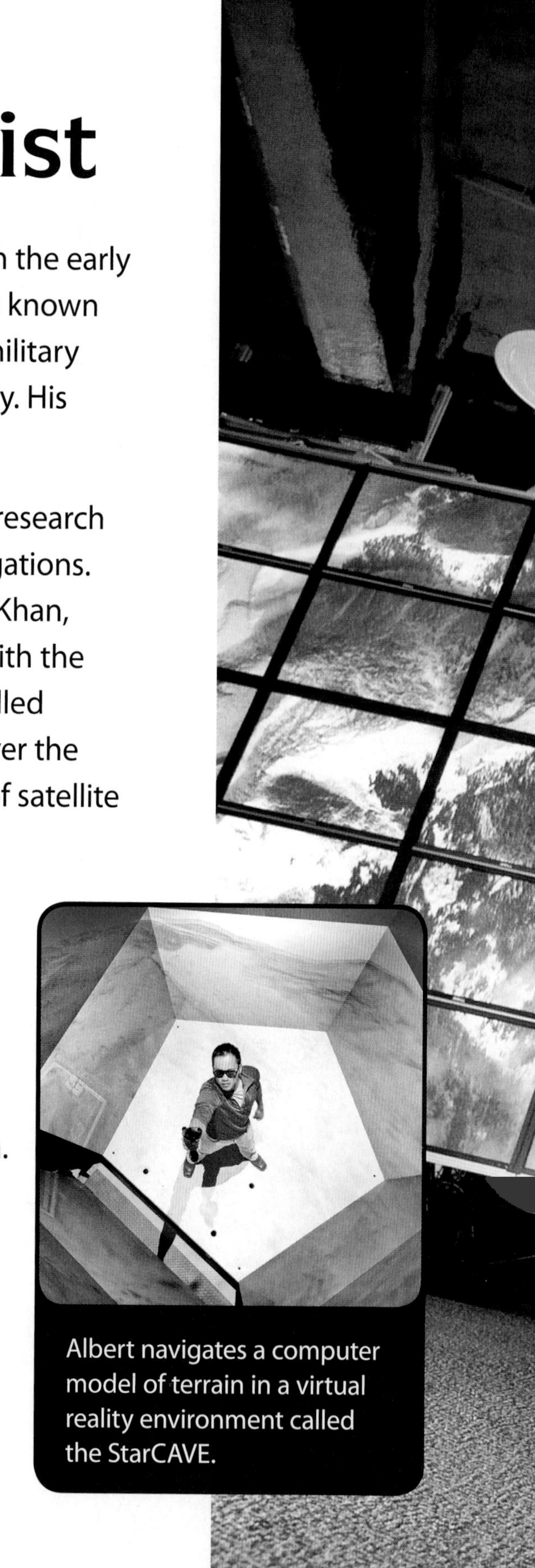

Albert navigates a computer model of terrain in a virtual reality environment called the StarCAVE.

NS Scientific Investigations Use a Variety of Methods. Science investigations use a variety of tools and techniques. (5-PS1-2)

NATIONAL GEOGRAPHIC | Explorer

Albert Yu-Min Lin is a research scientist at the University of California, San Diego. His quest for Genghis Khan's tomb is featured in a documentary called *The Forbidden Tomb of Genghis Khan* and has taken him to some of the most isolated areas in the world. Albert also enjoys public speaking, mountain climbing, surfing, and photography.

The wall of computer screens behind Albert displays the most detailed satellite images available. The display shows part of Mongolia, a very challenging place to explore.

Check In

My Science Notebook

Congratulations! You have completed *Physical Science*. Now let's reflect on what you have learned. Here you'll find a checklist and some additional questions to help you assess your progress. Page through your science notebook to find examples of different items. On a separate page in your science notebook, write your assessment of your work so far.

Read each item in this list. Ask yourself if you think you did a good job of it.

For each item, select the choice that is true for you: A. Yes **B.** Not Yet

- I defined and illustrated science vocabulary and main ideas.
- I labeled drawings. I wrote notes to explain ideas.
- I collected photos, news stories, and other objects.
- I used tables, charts, or graphs to record and analyze data.
- I included evidence for explanations and conclusions.
- I described how scientists and engineers answer questions and solve problems.
- I asked new questions of my own.
- I did something else. (Describe it.)

Reflect on Your Learning

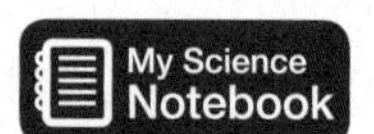

1. Which investigation interested you the most? Explain why you liked it.
2. Choose one science idea that you would like to learn more about. What questions can you ask?

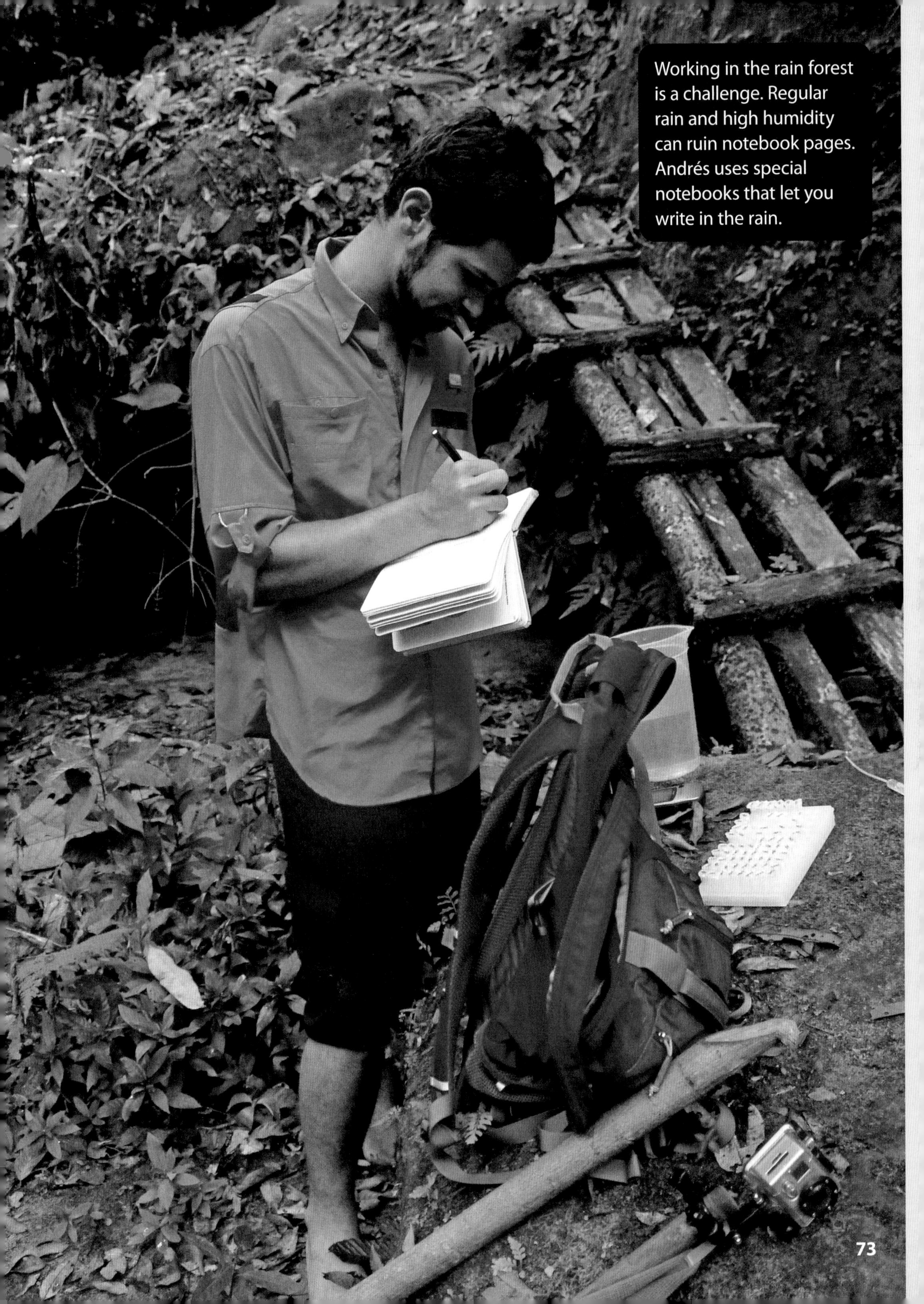

Working in the rain forest is a challenge. Regular rain and high humidity can ruin notebook pages. Andrés uses special notebooks that let you write in the rain.

Explorer

Andrés Ruzo Geoscientist
National Geographic Explorer

Let's Explore!

Through science, people continually improve their understanding of the natural world. Scientists form the best explanations they can by collecting and analyzing data and then making inferences to make conclusions that are based on evidence. Scientists use new evidence to refine scientific explanations. Both the explanations and our understanding keep growing! And new discoveries lead to new questions.

Life science is the study of living things and their environments. While studying the Boiling River, I note the local wildlife and how they are affected by the river. Here are some questions you might ask about *Life Science*:

- Where did the energy stored in your lunch originally come from?
- Why are fungi and bacteria so important in nature?
- How can scientists and engineers solve the problem of growing crops when water is scarce and the soil is low in nutrients?
- How did a plant get the nickname "the vine that ate the South"?
- What surprising thing do certain small wasps do to a species of red imported fire ants? And why is this a good thing?

Look at the drawings for some notebook examples and ideas. As you read, think of questions of your own. Then let's check in again to compare notes!

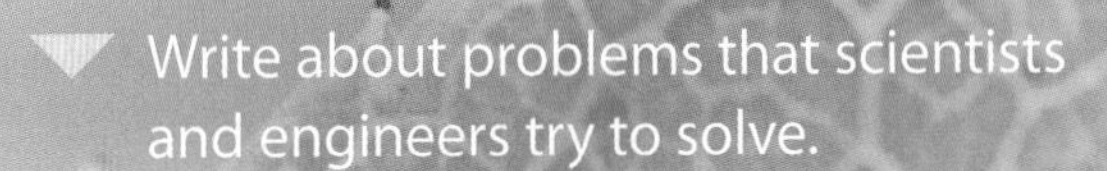

Write about problems that scientists and engineers try to solve.

Problem: The soil is dry and low in nutrients. The crops are damaged by insects and other pests.

Write about solutions that scientists and engineers find to solve problems.

Solution: Hydroponics—plants can grow without soil. Nutrients are added to the water in the pipes. The crops now have enough water and nutrients to grow.

In your science notebook, use drawings and write notes to explain main ideas.

Meadow Food Web

A food web shows how energy, originally from the sun, flows among living things in the environment.

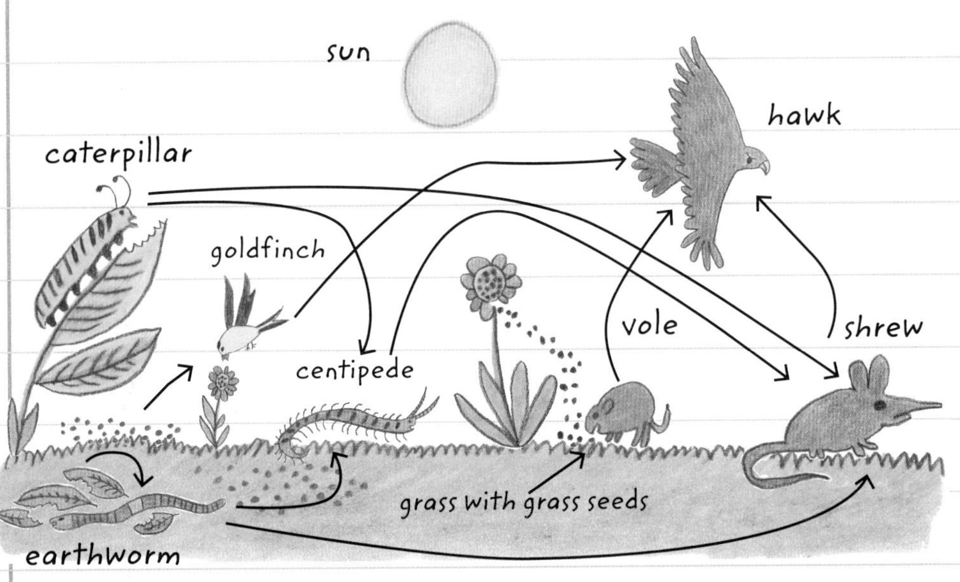

Life Science

Matter and Energy in Organisms and Ecosystems

The Blue-fronted Amazon parrot feeds on fruit in a tropical rain forest in South America.

What Plants Need

In the treetops of a rain forest, an orchid plant clings to the branch of a huge tree. The orchid and the tree are very different. Yet both of these plants need three main things to grow—sunlight, air, and water.

Like all living things, plants require food for energy. But unlike animals, plants are able to make their own food. They do this by using the energy of sunlight.

The orchid gets water from rain and fog.

In a rain forest, the tall trees compete for sunlight. Their leaves block much of the light. Only a few kinds of plants can grow in the deep shade of the forest floor. Many small plants, such as this orchid, grow on the tall trees' branches. By growing on another plant, the orchid is able to get the sunlight it needs to survive.

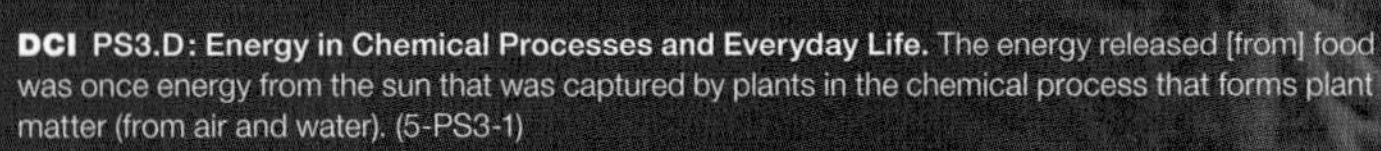

DCI PS3.D: Energy in Chemical Processes and Everyday Life. The energy released [from] food was once energy from the sun that was captured by plants in the chemical process that forms plant matter (from air and water). (5-PS3-1)

CCC Energy and Matter. Energy can be transferred in various ways and between objects. (5-PS3-1)

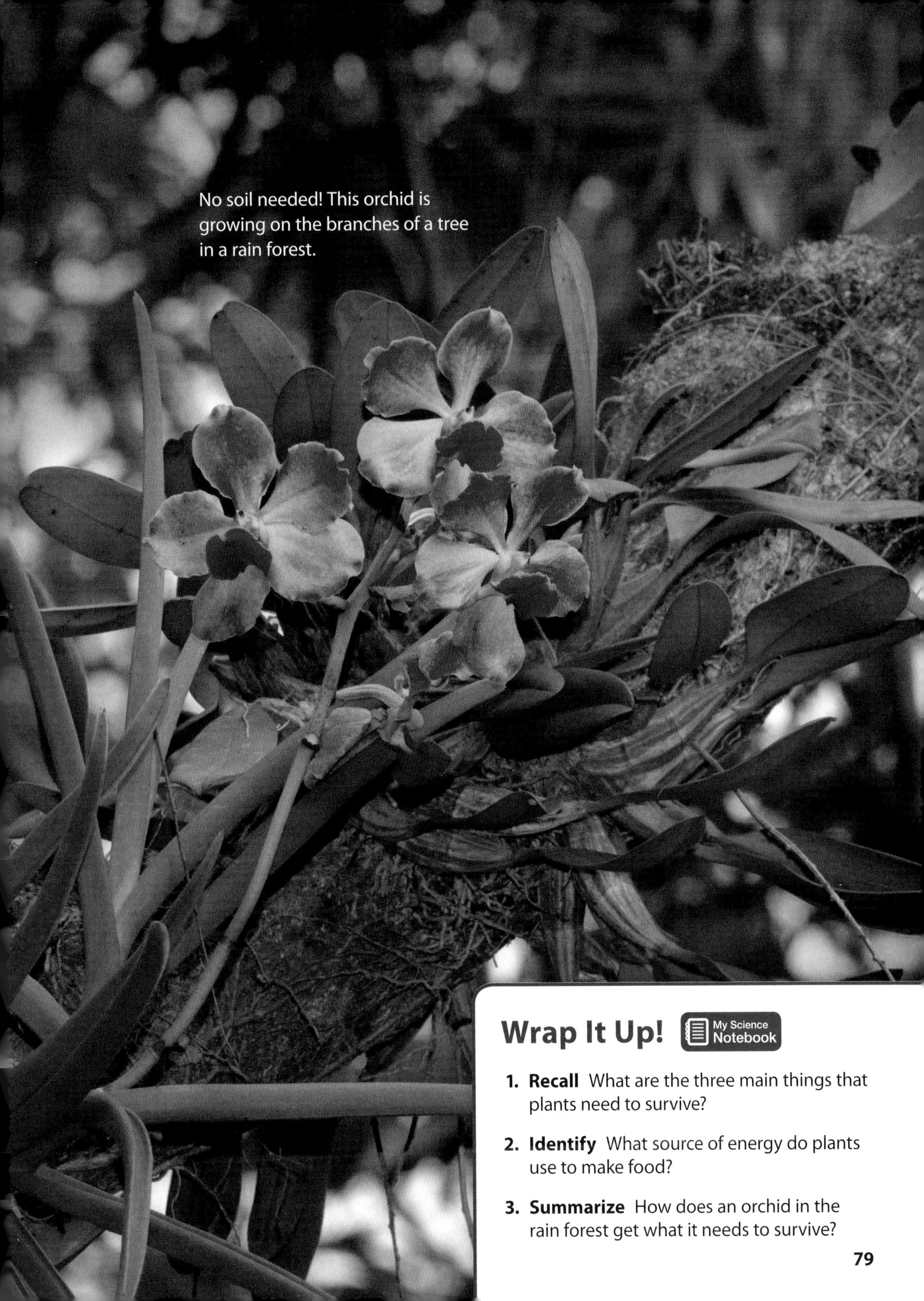

No soil needed! This orchid is growing on the branches of a tree in a rain forest.

Wrap It Up!

1. **Recall** What are the three main things that plants need to survive?
2. **Identify** What source of energy do plants use to make food?
3. **Summarize** How does an orchid in the rain forest get what it needs to survive?

How Plants Get Energy

Plants make their own food in a chemical process called **photosynthesis.** Photosynthesis takes place in leaves, where the green pigment **chlorophyll** captures the energy of sunlight. Leaves use this energy to change carbon dioxide and water into sugar. The carbon dioxide comes from the air. The water is taken up from soil or from the air.

The main product of photosynthesis is food in the form of sugar. Sugar stores energy. Plants use the energy released from sugar to carry out their basic functions. So the energy released from sugar was once energy from the sun. Photosynthesis also produces oxygen, which is released into the air. The diagram shows the chemical process of photosynthesis.

Sunlight provides the energy that plants use to produce their food.

DCI PS3.D: Energy in Chemical Processes and Everyday Life. The energy released [from] food was once energy from the sun that was captured by plants in the chemical process that forms plant matter (from air and water). (5-PS3-1)

CCC Energy and Matter. Energy can be transferred in various ways and between objects. (5-PS3-1)

Wrap It Up! My Science Notebook

1. **Explain** What happens during the process of photosynthesis?
2. **Identify** What substance allows plants to capture the sun's energy?
3. **Summarize** What two materials are used in photosynthesis? Where does each of them come from?

Materials for Plant Growth

Wild bamboo plants grow very quickly. Some can grow more than 60 centimeters (2 feet) in a single day!

Where does the material for this amazing growth come from? You may think it comes from soil, but it does not. Most of it comes from carbon dioxide, a gas in the air, and water. Remember that, during photosynthesis, plants capture energy from the sun and produce sugar. Plants use that sugar as a source of stored energy. Plants also use sugar as a building block to make new leaves, stems, and roots.

In addition to air and water, plants need mineral **nutrients,** which include nitrogen, calcium, and magnesium. Plants use these nutrients to make important substances such as proteins and DNA.

Mineral nutrients are dissolved in water. As a plant takes in water, it also takes in nutrients. When gardeners and farmers add fertilizers to soil or water, they are adding mineral nutrients.

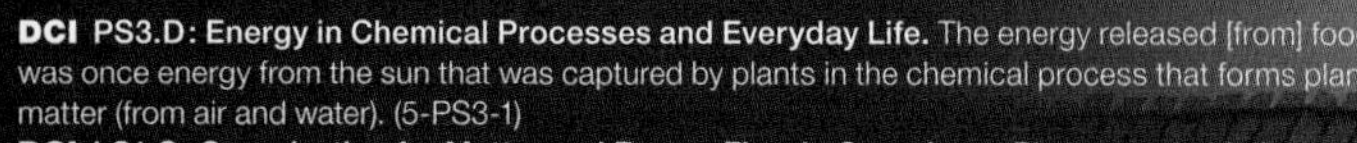
DCI PS3.D: Energy in Chemical Processes and Everyday Life. The energy released [from] food was once energy from the sun that was captured by plants in the chemical process that forms plant matter (from air and water). (5-PS3-1)
DCI LS1.C: Organization for Matter and Energy Flow in Organisms. Plants acquire their material for growth chiefly from air and water. (5-LS1-1)
CCC Energy and Matter. Energy can be transferred in various ways and between objects. (5-PS3-1)

Bamboo grows very quickly. It gets the materials for growth from air, water, and dissolved nutrients.

The nutrients these bamboo plants need are dissolved in the water in the vase. The bamboo does not need soil to live and grow.

Wrap It Up! My Science Notebook

1. **Recall** Where does most of the material in a plant come from?
2. **Describe** Where do plants get mineral nutrients?
3. **Analyze** Sometimes people call fertilizers "plant food." Is this an accurate way to describe mineral nutrients? *Hint*: Do mineral nutrients provide a plant with energy?

Growing Crops

Insects and other pests may devour crops, leaving little food for humans. Raccoons damaged this sweet corn.

Problem

How can people grow crops where there is not enough land with good soil?

Nearly all plants grow in soil. Some places on Earth have thick, fertile soil that supports crop growth. But in many parts of the world, there is a shortage of land with good soil. Even in places where the soil is rich, other problems such as pests and too much or too little rain can harm crops.

Why is crop growth important? Today there are about 7 billion people in the world. By 2050 that number is expected to increase to more than 9 billion. All those people will need much more food. As the population grows, more and more land will be used for cities, roads, and homes. Where will that food be grown?

DCI LS1.C: **Organization for Matter and Energy Flow in Organisms.** Plants acquire their material for growth chiefly from air and water. (5-LS1-1)
CCC **Energy and Matter.** Matter is transported into, out of, and within systems. (5-LS1-1)

In low-lying areas, frequent floods can destroy crops.

Some crops are damaged when rain does not fall regularly. Severe lack of rain is called drought.

In some parts of the world, the soil lacks nutrients or is too dry for crops to grow.

Solution

One solution is hydroponics. In **hydroponics,** plants are grown in water instead of soil. Nutrients plants would normally get from water in soil are added to the water in which the plants grow. Because these systems are often set up in greenhouses, fresh local food is made available year-round.

You might think that hydroponic systems would use more water than traditional methods of farming. But hydroponics can actually save water, since the plants do not have to be irrigated. That makes hydroponics very useful in arid regions. Plants still get sunlight and air they need. In parts of Africa, scientists have found that crops grown with hydroponics use only one third of the water used by a traditional farm. In Arizona, some greenhouse hydroponic systems use only ten percent of the water used by a traditional system.

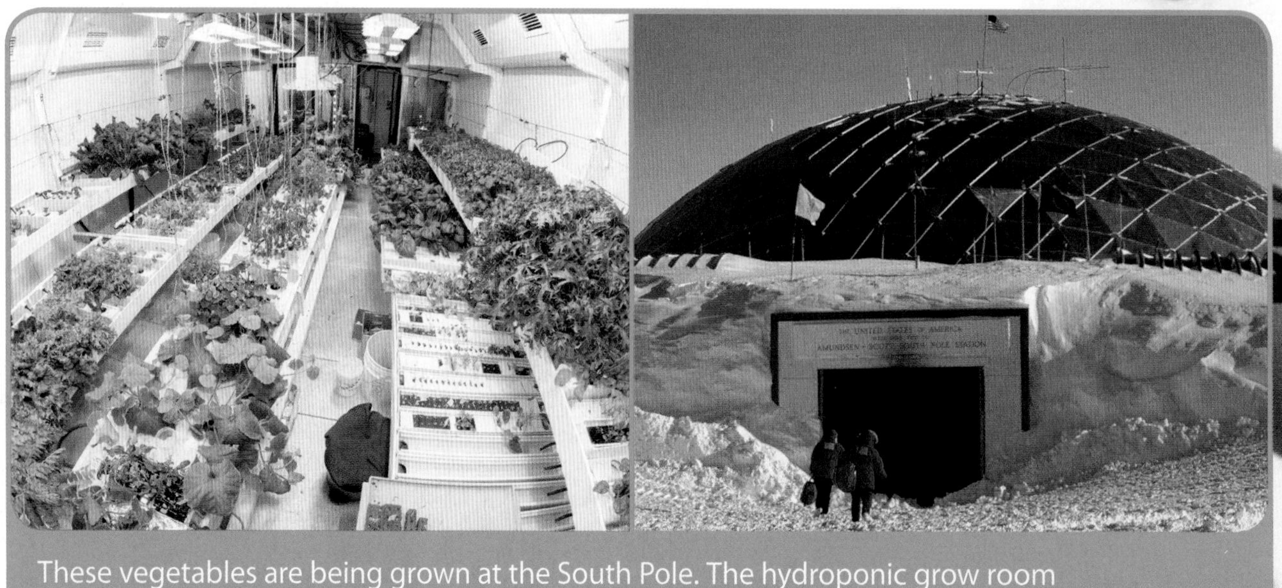

These vegetables are being grown at the South Pole. The hydroponic grow room produces enough vegetables to provide 50 people at least one salad a week.

Plants in hydroponic systems are often grown in racks so that they take up little space.

Wrap It Up! My Science Notebook

1. **Define** What is hydroponics?
2. **Compare and Contrast** Compare the way that plants grown in soil and plants grown by hydroponics get the things they need. Include sunlight, water, air, and mineral nutrients.
3. **Apply** As the world population increases, there may be less land available to grow food. How can hydroponics help solve this problem?

Investigate

Hydroponics

How can you grow plants without soil?

If people travel long distances in space, how could they grow food for themselves? There isn't any soil in space. One solution would be to use hydroponics. Such a system could easily produce vegetables, such as tomatoes and lettuce. Imagine eating a fresh salad while traveling in space!

Does it seem amazing that astronauts can grow plants without soil? In this investigation, you will gather evidence to show that it is possible.

Materials

- clear plastic container and lid with a hole in the middle
- young plant
- water
- cotton
- liquid houseplant fertilizer

PE 5-LS1-1. Support an argument that plants get the materials they need for growth chiefly from air and water.

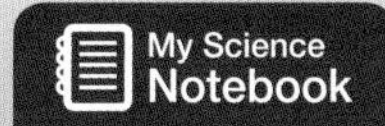

1 Pour water into the container until it is half full. Add 5 drops of liquid houseplant fertilizer to the water to provide nutrients for the plant.

2 Gently push the roots of the plant through the hole in the lid. Then wrap the stem with cotton. Attach the lid to the container.

3 Place the container in a sunny place, such as a windowsill. If necessary, add more water so the roots are covered. Observe and record any signs of plant growth, such as new leaves. Observe and record any changes in the water level.

4 Observe the plant for 5 or more days. Add water to the container to keep the roots under water. Record your observations.

Wrap It Up!

1. **Observe** What changes, if any, did you observe in the plant and the water level?
2. **Infer** What would happen if you put the plant in a dark closet? Why?
3. **Analyze** What are the main parts of a hydroponic system? How is matter, such as water and air, transported into, out of, and within the system?
4. **Draw Conclusions** Use evidence from your investigation to support an argument that plants can grow without soil.

Support an Argument

Orchids, bamboo plants, and corn plants are very different from each other, but they all need to take in materials for growth. Because most of the plants we see grow in soil, many people think plants use soil as a building block for their growth. How could you support an argument that describes where plants get the materials they need to live and grow? Your teacher will give you data from a plant growth experiment. The plants were grown in different conditions and different measurements were taken over time. Use this data and examples and evidence from previous lessons to decide whether plants need the following materials to live and grow:

- water
- soil
- carbon dioxide from air
- nutrients

1. **List.** 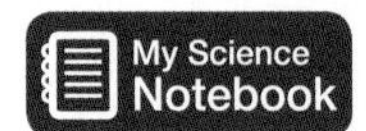

   Which materials did you decide that plants need to live and grow? What examples or evidence did you use to make your decision?

2. **Compare.**
   Work with a group. Compare your lists. Discuss your choices, and work together until you can agree about which materials are needed for plants to live and grow.

How are these plants getting the materials they need for growth?

PE 5-LS1-1. Support an argument that plants get the materials they need for growth chiefly from air and water.

3. **Construct an argument.**
   Have each person in the group choose one of the materials and explain why a plant must have that material to live and grow or why the plant does not need that material.

4. **Generalize.**
   Come back together as a group. Present your argument for whether plants need carbon dioxide from air, water, nutrients, or soil as building blocks to grow.

Why Animals Need Food

It's a misty morning in the Ngorongoro Crater of Tanzania. An African elephant has already walked several miles to reach a grove of acacia trees. Now the elephant reaches up with its long, flexible trunk and pulls the leaves of the tree to its mouth. The elephant will use its wide, flat teeth to grind the tough leaves. Then it will swallow the leaves and digest its breakfast. The elephant needs to eat the acacia leaves to get energy. But it must also use energy to move and to get its food.

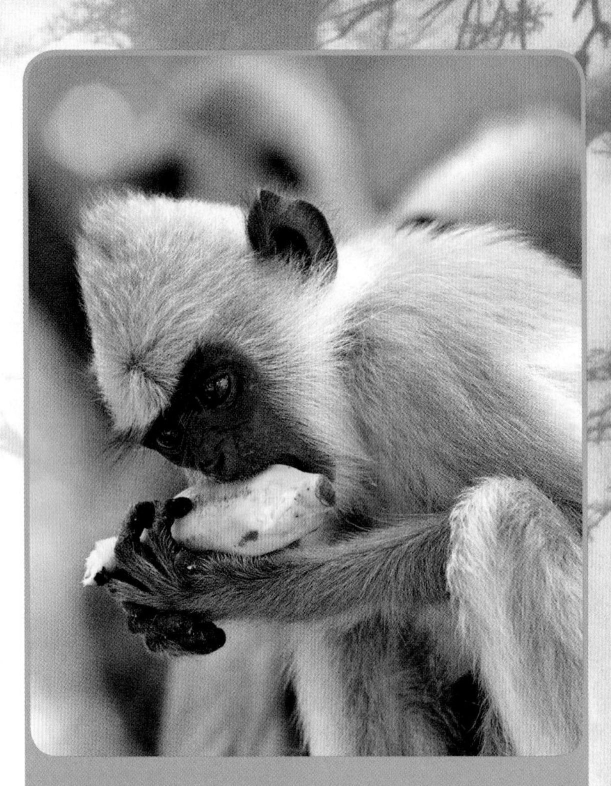

The Grey Langur monkey gets the energy it needs from the fruit it eats.

Like the elephant, all animals need to eat food. Food provides animals with the materials they need to grow larger and to repair their bodies. Food also provides animals with the energy they need for motion and to carry out other life functions, such as digestion. Warm-blooded animals, such as mammals and birds, use the energy in food to keep their bodies warm.

DCI **LS1.C: Organization for Matter and Energy Flow in Organisms.** Food provides animals with the materials they need for body repair and growth and the energy they need to maintain body warmth and for motion. (secondary to 5-PS3-1)

CCC **Energy and Matter.** Energy can be transferred in various ways and between objects. (5-PS3-1)

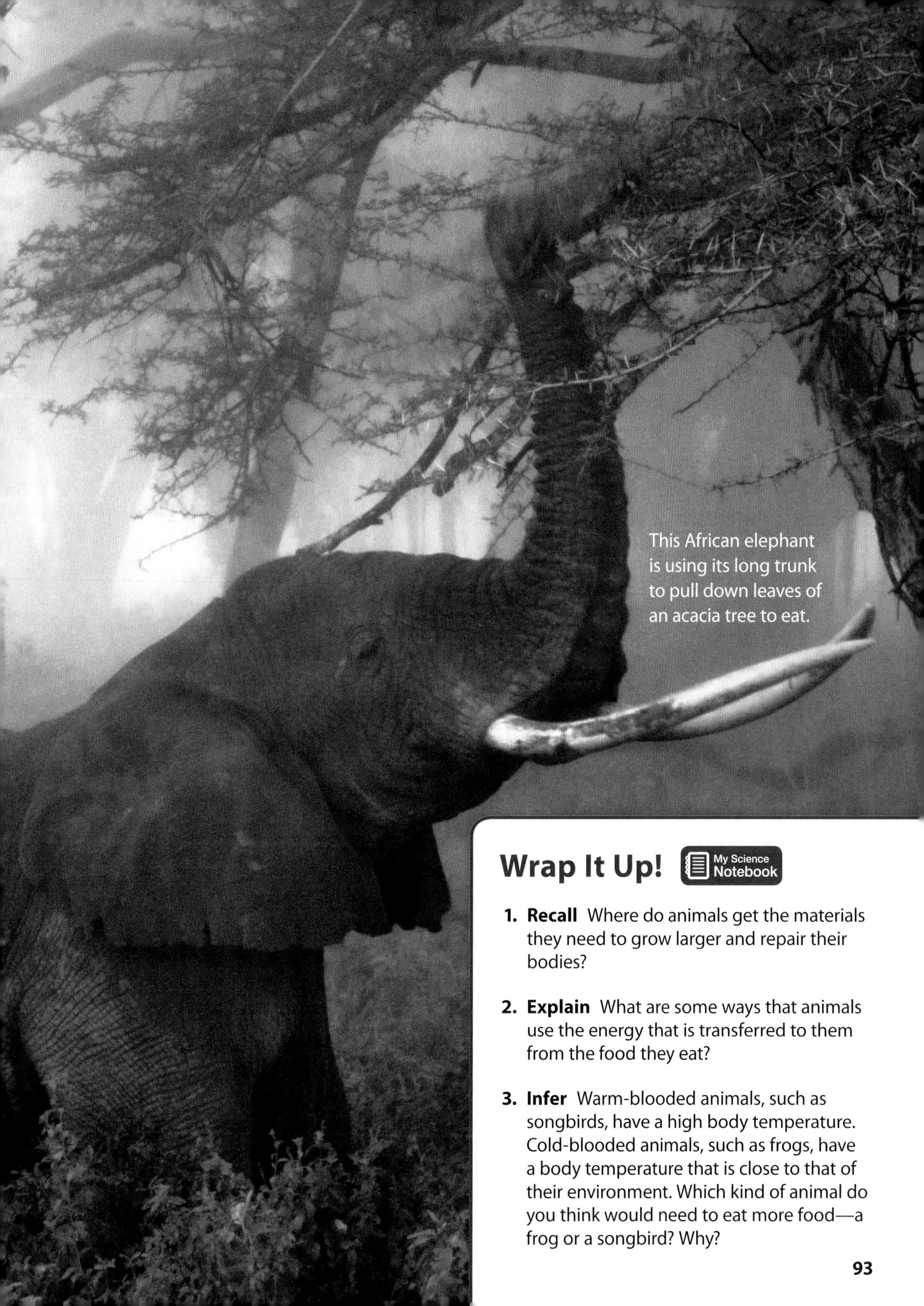

This African elephant is using its long trunk to pull down leaves of an acacia tree to eat.

Wrap It Up! My Science Notebook

1. **Recall** Where do animals get the materials they need to grow larger and repair their bodies?

2. **Explain** What are some ways that animals use the energy that is transferred to them from the food they eat?

3. **Infer** Warm-blooded animals, such as songbirds, have a high body temperature. Cold-blooded animals, such as frogs, have a body temperature that is close to that of their environment. Which kind of animal do you think would need to eat more food—a frog or a songbird? Why?

Desert Food Chain

Every organism needs energy to survive. Organisms that make their own food, such as plants, are called **producers.** Producers get their energy directly from the sun. **Consumers** get energy by eating other

Energy Source
Sunlight is the source of energy for the food chain.

Producer
Producers use energy from sunlight to make their own food.

Consumer
Some consumers get energy by eating producers.

DCI LS1.C: Organization for Matter and Energy Flow in Organisms. Food provides animals with the materials they need for body repair and growth and the energy they need to maintain body warmth and for motion. (secondary to 5-PS3-1)
DCI PS3.D: Energy in Chemical Processes and Everyday Life. The energy released [from] food was once energy from the sun that was captured by plants in the chemical process that forms plant matter (from air and water). (5-PS3-1)
CCC Energy and Matter. Energy can be transferred in various ways and between objects. (5-PS3-1)

organisms. Many consumers eat plants, but other consumers eat animals. Still other consumers eat both plants and animals.

A food chain is a path by which energy flows from one living thing to another in an environment. The photos below show a food chain in a desert. The arrows show one way that energy moves from one living thing to another. Use your finger to trace the path of energy.

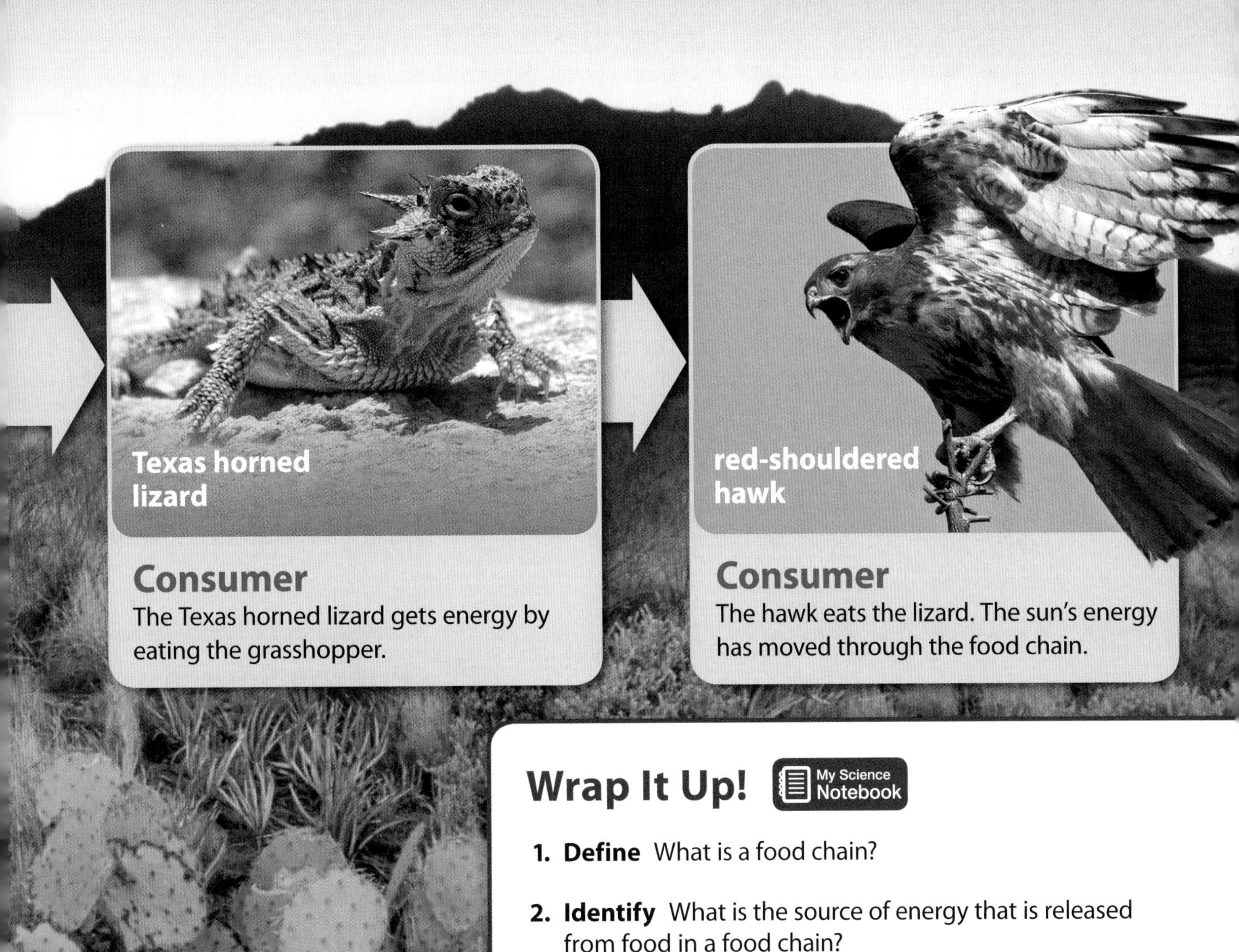

Consumer
The Texas horned lizard gets energy by eating the grasshopper.

Consumer
The hawk eats the lizard. The sun's energy has moved through the food chain.

Wrap It Up!

My Science Notebook

1. **Define** What is a food chain?
2. **Identify** What is the source of energy that is released from food in a food chain?
3. **Compare** How do producers and consumers each obtain the energy they need to live and grow?
4. **Analyze** Could producers live without consumers? Could consumers live without producers? Explain.

Compare and Contrast

The diagrams you see here show food chains in two different environments. One food chain is in a pond and the other is in a rain forest.

In any food chain, the energy contained in food was once energy from the sun. In a rainforest food chain, plants, such as trees, capture the sun's energy in chemical processes that form plant matter. Tiny, floating algae in a pond food chain are not plants, but like plants, they are producers. They use energy from the sun to produce their own food.

Compare the two food chains shown here. How are they alike? How are they different?

Energy Source
Sunlight is the source of energy for the food chain.

Energy Source
Sunlight is the source of energy for the food chain.

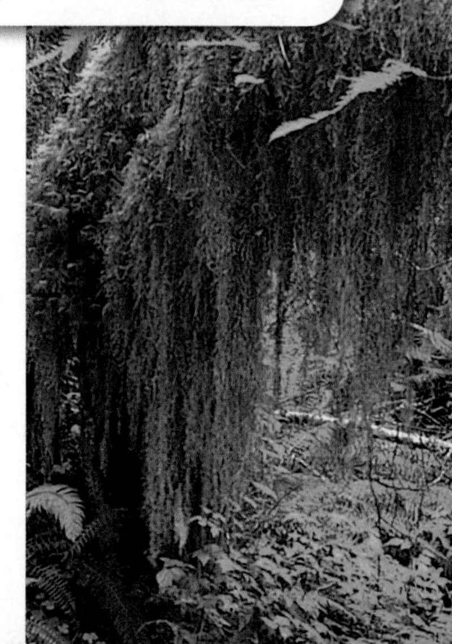

DCI **LS1.C: Organization for Matter and Energy Flow in Organisms.** Food provides animals with the materials they need for body repair and growth and the energy they need to maintain body warmth and for motion. (secondary to 5-PS3-1)
DCI **PS3.D: Energy in Chemical Processes and Everyday Life.** The energy released [from] food was once energy from the sun that was captured by plants in the chemical process that forms plant matter (from air and water). (5-PS3-1)
CCC **Energy and Matter.** Energy can be transferred in various ways and between objects. (5-PS3-1)

POND FOOD CHAIN

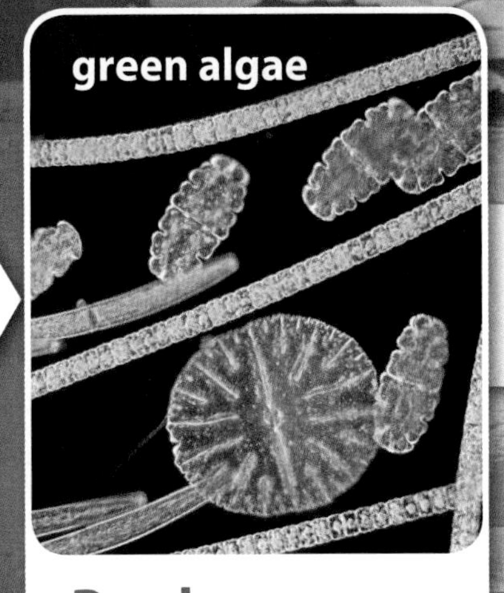

Producer
Algae use sunlight to grow and reproduce.

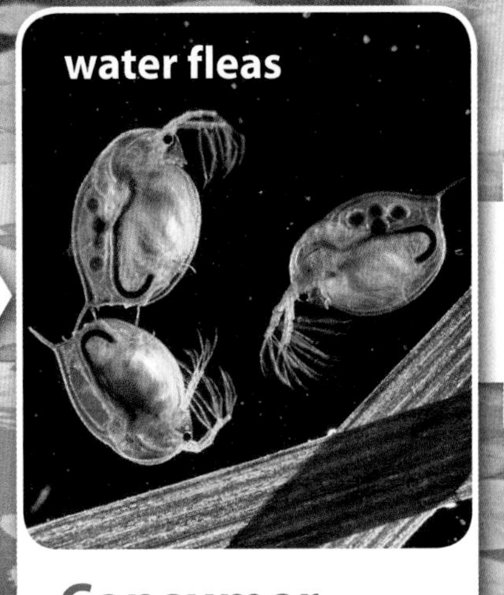

Consumer
Tiny animals such as water fleas eat algae.

Consumer
Small fish eat smaller animals such as water fleas.

RAIN FOREST FOOD CHAIN

Producer
Rain forest trees use sunlight to make food. They use the energy in this food to produce fruit and seeds.

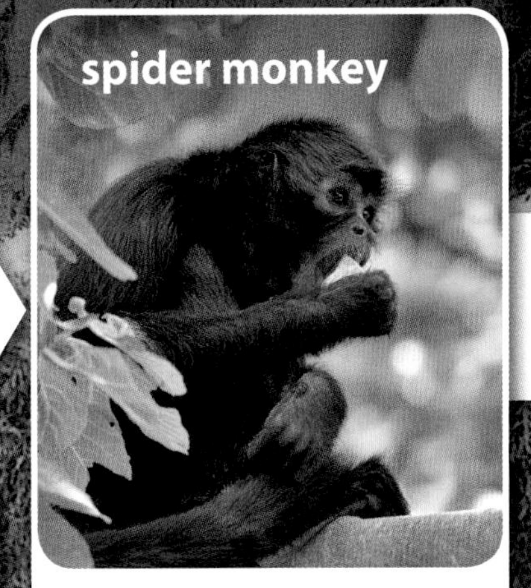

Consumer
Monkeys eat fruits and seeds.

Consumer
Jaguars eat animals, such as monkeys.

Wrap It Up!

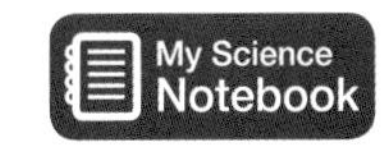

1. **Interpret Diagrams** What is the original source of energy for both food chains?
2. **Compare and Contrast** How are the producers and consumers in the pond like those in the rain forest? How are they different?

Use Models

You've learned that all animals need energy to grow, move, and repair their bodies. You've also seen some of the ways animals get this energy from their environment through food chains. Now it's your turn. Using your own research, you will make and present a model food chain for an environment. The food chain should include at least two different kinds of animals.

1. **Ask a question.** 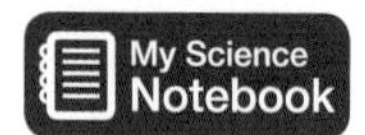

   How can you use a model to describe that energy in animals' food was once energy from the sun?

2. **Research an environment.**
   First pick an environment that you would like to learn about. Then use print and online reference materials to discover what kinds of plants and animals live in that environment. Be sure to record in your science notebook the source of all the information you find.

   After you have some information, think about the organisms you have identified. Which are the producers? What consumers eat those producers? And what consumers eat those consumers? Identify one or two pathways that energy takes through the environment.

PE 5-PS3-1. Use models to describe that energy in animals' food (used for body repair, growth, motion, and to maintain body warmth) was once energy from the sun.

3. **Assemble your model.**
   How will you make your model food chain? You might use index cards, clay models, or computer slide-show presentation software. Include labels for each step in the chain. Your labels should give a brief description of how each organism gets energy.

4. **Analyze and revise your model.**
   Do research to find another plant or animal from your selected environment. Add this organism to the model food chain.

5. **Present your model.**
   When you have revised your model, present your model food chain to the class. Point out how each organism in the food chain gets energy for life processes. Then use your model to explain how all food energy was once energy from the sun.

The sun is the original source of energy in every food chain on the African savanna.

Desert Food Web

A rattlesnake in the desert eats animals such as mice and grasshoppers. These animals are part of the desert **food web.** A food web is a combination of food chains that shows how energy moves from the sun through an environment. Food webs show that consumers get energy from a variety of organisms.

plants

DCI LS2.A: Interdependent Relationships in Ecosystems. The food of almost any kind of animal can be traced back to plants. Organisms are related in food webs in which some animals eat plants for food and other animals eat the animals that eat plants. Some organisms, such as fungi and bacteria, break down dead organisms (both plants or plants parts and animals) and therefore operate as "decomposers." Decomposition eventually restores (recycles) some materials back to the soil. Organisms can survive only in environments in which their particular needs are met. A healthy ecosystem is one in which multiple species of different types are each able to meet their needs in a relatively stable web of life. Newly introduced species can damage the balance of an ecosystem. (5-LS2-1)

CCC Systems and System Models. A system can be described in terms of its components and their interactions. (5-LS2-1)

Trace the Energy! The arrows show the direction in which energy moves through the food web. For example, plants get energy from sunlight. The bighorn sheep gets energy from plants. The cougar gets energy from the bighorn sheep. Use your finger to trace the energy through different food chains in the desert food web.

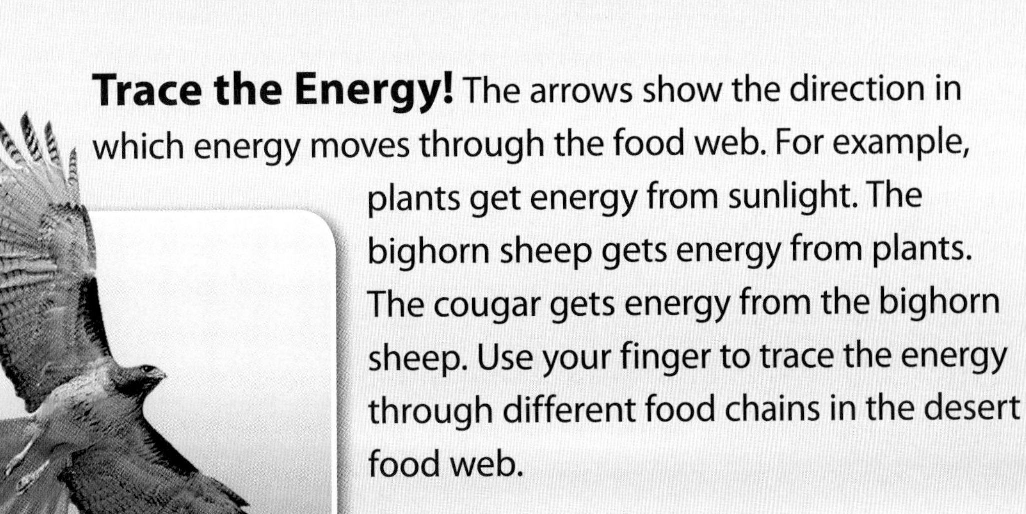

Wrap It Up! My Science Notebook

1. **Explain** How does energy flow through a food chain?

2. **Contrast** How is a food web different from a food chain?

3. **Infer** Suppose a disease kills most of the hawks in a part of the desert. How might the loss of the hawks affect the other animals in the area?

plants. The mushroom is a **decomposer,** an organism that gets its energy by breaking down dead organisms or the waste of living things. **Fungi,** such as the mushroom, and microscopic organisms called **bacteria** are two kinds of decomposers.

When any plant or animal in a food web dies, it is broken down by decomposers. Decomposition eventually restores, or recycles, nutrients back to the soil. The nutrients released by decomposers help plants grow.

These red Amanita mushrooms get energy by breaking down leaves and other plant parts that fall onto the forest floor.

DCI LS2.A: Interdependent Relationships in Ecosystems. The food of almost any kind of animal can be traced back to plants. Organisms are related in food webs in which some animals eat plants for food and other animals eat the animals that eat plants. Some organisms, such as fungi and bacteria, break down dead organisms (both plants or plants parts and animals) and therefore operate as "decomposers." Decomposition eventually restores (recycles) some materials back to the soil. Organisms can survive only in environments in which their particular needs are met. A healthy ecosystem is one in which multiple species of different types are each able to meet their needs in a relatively stable web of life. Newly introduced species can damage the balance of an ecosystem. (5-LS2-1)

CCC Systems and System Models. A system can be described in terms of its components and their interactions. (5-LS2-1)

These bacteria are decomposers that get their energy from dead leaves. The image is greatly magnified.

Wrap It Up! My Science Notebook

1. **Name** What are two kinds of decomposers?
2. **Explain** How do decomposers get energy?
3. **Cause and Effect** Suppose there were no decomposers in the soil. How might this affect plants growing in the area?

Cycles of Matter

The grizzly bear walking among the grasses in the meadow may look like he is alone, but he's not! Other animals such as rabbits, mice, and deer live here. Fungi and bacteria, as well as insects and earthworms, live in the soil. All of these organisms are connected by cycles of matter. These cycles also include the air and soil.

All of the organisms in the meadow play a part in the carbon dioxide-oxygen cycle.

Carbon Dioxide The grizzly bear breathes in oxygen. It uses the oxygen to break down and get energy from food. The bear gives off carbon dioxide as waste. Plants use carbon dioxide to make food.

DCI LS2.B: Cycles of Matter and Energy Transfer in Ecosystems. Matter cycles between the air and soil and among plants, animals, and microbes as these organisms live and die. Organisms obtain gases, and water, from the environment, and release waste matter (gas, liquid, or solid) back into the environment. (5-LS2-1)

CCC Systems and System Models. A system can be described in terms of its components and their interactions. (5-LS2-1)

Carbon dioxide and oxygen are gases found in the air. The carbon dioxide-oxygen cycle provides living things with the carbon and oxygen they need to survive.

Organisms in the meadow are also connected by the nitrogen cycle. Nitrogen is a substance found in the air and in the soil. Plants take in nitrogen from the soil. Animals get the nitrogen they need by eating plants or other animals. When plants and animals die, decomposers return the nitrogen in their bodies to the soil. Plant and animal wastes also contain nitrogen, which microbes return to the soil.

Oxygen When plants make their own food through photosynthesis, they take in carbon dioxide from the air and give off oxygen as waste. They also use oxygen to break down and get energy from food.

Wrap It Up!

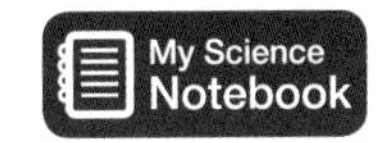

1. **Describe** What is the role of decomposers in the nitrogen cycle?
2. **Explain** Why is the carbon dioxide-oxygen cycle important to plants and animals?
3. **Sequence** The following organisms are part of the nitrogen cycle: microscopic decomposers, plant, rabbit. Draw a diagram with arrows that puts the organisms in the correct order. Begin with nitrogen in the soil.

Tallgrass Prairie Ecosystem

These bison live in the tallgrass prairie of North America. All of the plants and animals in a tallgrass prairie interact with the nonliving things in their environment. Some of these things are physical characteristics, such as the summer and winter temperatures, amount of rainfall, and kind of soil. Grasses grow in the deep, fertile soil. Animals drink water and breathe the air. These organisms also interact with the many other living things in the prairie.

These bison live in the Tallgrass Prairie Reserve in Oklahoma.

DCI LS2.A: Interdependent Relationships in Ecosystems. The food of almost any kind of animal can be traced back to plants. Organisms are related in food webs in which some animals eat plants for food and other animals eat the animals that eat plants. Some organisms, such as fungi and bacteria, break down dead organisms (both plants or plants parts and animals) and therefore operate as "decomposers." Decomposition eventually restores (recycles) some materials back to the soil. Organisms can survive only in environments in which their particular needs are met. A healthy ecosystem is one in which multiple species of different types are each able to meet their needs in a relatively stable web of life. Newly introduced species can damage the balance of an ecosystem. (5-LS2-1)

CCC Systems and System Models. A system can be described in terms of its components and their interactions. (5-LS2-1), (5-ESS2-1), (5-ESS3-1)

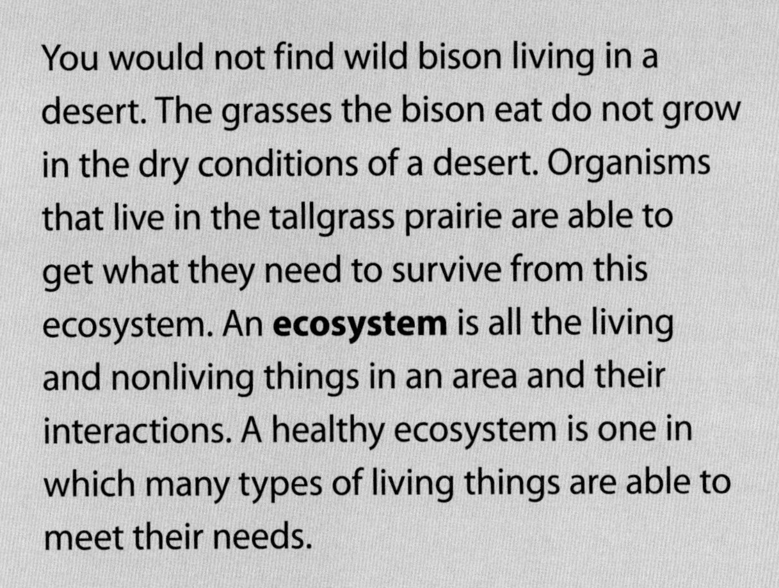

You would not find wild bison living in a desert. The grasses the bison eat do not grow in the dry conditions of a desert. Organisms that live in the tallgrass prairie are able to get what they need to survive from this ecosystem. An **ecosystem** is all the living and nonliving things in an area and their interactions. A healthy ecosystem is one in which many types of living things are able to meet their needs.

Horned larks live in the tallgrass prairie. They weave their nests from the fine grasses. Adults feed on the seeds of grasses and wildflowers. They feed insects to their young.

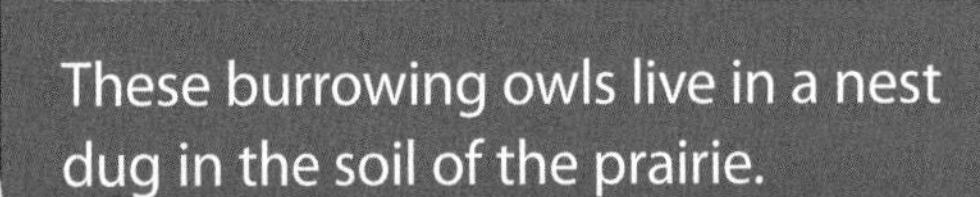

These burrowing owls live in a nest dug in the soil of the prairie.

Wrap It Up!

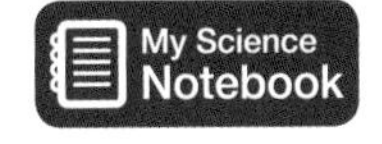

1. **Define** What is an ecosystem?
2. **Infer** What are some of the nonliving things you can observe or infer in this photo of a tallgrass prairie?
3. **Explain** How do the physical characteristics of an environment help support the organisms that live there?

Grassland Populations and Communities

Many different kinds of organisms, or **species,** live in the prairie ecosystem. Scientists classify the organisms in an ecosystem into three levels—individual organisms, populations, and communities. Individual prairie dogs usually live together in prairie dog towns. A **population** is all the individuals of a species living together in a particular place.

Individual
A single organism, such as this prairie dog, is an individual in an ecosystem.

Population
All the prairie dogs that live in a particular part of the ecosystem are a population.

DCI **LS2.A: Interdependent Relationships in Ecosystems.** The food of almost any kind of animal can be traced back to plants. Organisms are related in food webs in which some animals eat plants for food and other animals eat the animals that eat plants. Some organisms, such as fungi and bacteria, break down dead organisms (both plants or plants parts and animals) and therefore operate as "decomposers." Decomposition eventually restores (recycles) some materials back to the soil. Organisms can survive only in environments in which their particular needs are met. A healthy ecosystem is one in which multiple species of different types are each able to meet their needs in a relatively stable web of life. Newly introduced species can damage the balance of an ecosystem. (5-LS2-1)

CCC **Systems and System Models.** A system can be described in terms of its components and their interactions. (5-LS2-1)

Prairie dogs live in burrows, which they dig in the ground. Other animals, such as burrowing owls and black-footed ferrets, share their burrows. Prairie dogs eat the grasses and wildflowers that live in the prairie. They also serve as food for other animals, such as black-footed ferrets, foxes, and hawks. All of the populations that live and interact in an area make up a **community.** Organisms in a community can only survive in environments in which their particular needs are met. Healthy communities have many species connected by a variety of food webs.

Community
All the populations of organisms that live and interact in that part of the prairie form a community.

Ecosystem
All of the communities plus the physical parts of the environment that interact together make up the ecosystem.

Wrap It Up!

1. **List** What are the three levels of organisms that make up an ecosystem?

2. **Compare** How is a population different from a community?

Investigate

Interactions in a Model Pond

How do living things in a model pond ecosystem interact?

An ecosystem is all the living and nonliving elements in an area. You can make a model pond ecosystem and explore some of the interactions that take place there.

Materials

clear plastic bottle	sand	small rocks	Elodea
water	**spoon**	**3 snails**	**hand lens**

DCI **PS3.D: Energy in Chemical Processes and Everyday Life.** The energy released [from] food was once energy from the sun that was captured by plants in the chemical process that forms plant matter (from air and water). (5-PS3-1)

1 Layer sand and rocks at the bottom of a clear plastic bottle. Plant the Elodea. Pour water into the bottle until it is about two-thirds full. Use a spoon to place 3 snails in the model ecosystem.

2 Put your model in a sunny place. Observe the model each day for 7 days. Record your observations in your science notebook. Use a hand lens to observe changes in the ecosystem.

3 Use your observations to infer what each living thing needs and how it meets those needs. Classify each organism as a producer or consumer.

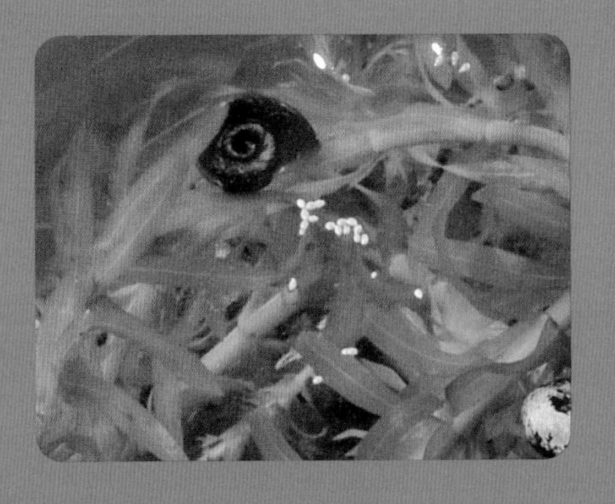

4 Draw your model pond ecosystem. Label the organisms. Draw arrows to show how energy moved from the sun to the producers to the consumers in the model. Then use the Internet or other resources to research one more producer and one more consumer that could be added to your model pond ecosystem. Revise your drawings of the model ecosystem based on the addition of these organisms.

Wrap It Up!

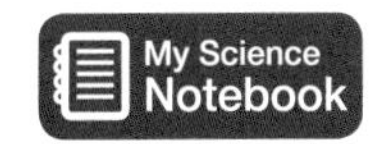

1. **Classify** How did your observations help you classify producers and consumers in your ecosystem?

2. **Compare and Contrast** In what ways is your model like a real pond? In what ways is it different?

Develop a Model

Ecosystems are made up of the organisms and nonliving things in the environment. Materials are constantly moving among these different parts. You can develop a model to describe this movement of matter.

1. **Ask a question.** 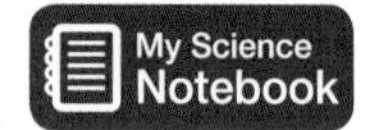

   How can you develop a model to describe the movement of matter among plants, animals, decomposers, and the environment?

2. **Research an ecosystem.**
   Select an ecosystem for your model. Will it be a small ecosystem, such as a pond, or a big ecosystem, such as a pine forest? Use reliable reference materials to discover how materials cycle through that ecosystem.

   Decide which cycle to show. Where do the raw materials for the cycle come from? Are they in the air, water, or soil? How are the materials taken in by living things? How do the materials move among different organisms? What happens to the materials when organisms die? Record all your findings in your science notebook.

PE 5-LS2-1. Develop a model to describe the movement of matter among plants, animals, decomposers, and the environment.

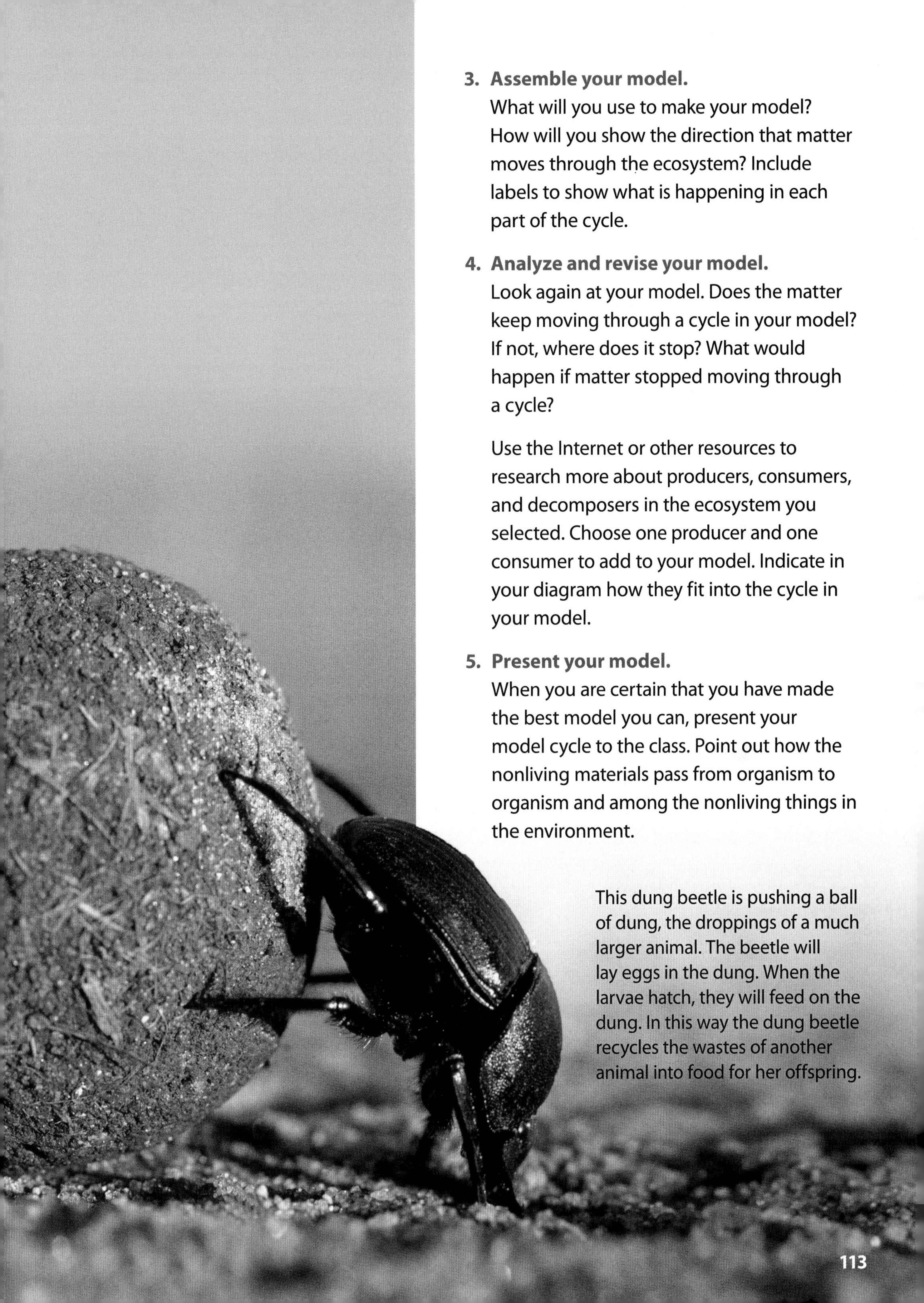

3. **Assemble your model.**
   What will you use to make your model? How will you show the direction that matter moves through the ecosystem? Include labels to show what is happening in each part of the cycle.

4. **Analyze and revise your model.**
   Look again at your model. Does the matter keep moving through a cycle in your model? If not, where does it stop? What would happen if matter stopped moving through a cycle?

   Use the Internet or other resources to research more about producers, consumers, and decomposers in the ecosystem you selected. Choose one producer and one consumer to add to your model. Indicate in your diagram how they fit into the cycle in your model.

5. **Present your model.**
   When you are certain that you have made the best model you can, present your model cycle to the class. Point out how the nonliving materials pass from organism to organism and among the nonliving things in the environment.

This dung beetle is pushing a ball of dung, the droppings of a much larger animal. The beetle will lay eggs in the dung. When the larvae hatch, they will feed on the dung. In this way the dung beetle recycles the wastes of another animal into food for her offspring.

STEM

SCIENCE
TECHNOLOGY
ENGINEERING
MATH

ENGINEERING PROJECT

Design an Aquaponics System

Have you ever seen a fish farm? In aquaculture, people raise fish in ponds and harvest them for food. But the fish also produce a lot of waste, which must be removed from the ponds. Recently, people have found a way to manage this waste by combining aquaculture with hydroponics. This combination is called aquaponics.

In aquaponics, wastewater from fishponds is circulated around the roots of plants. Nutrients, such as nitrogen in the wastewater, fertilize the plants and help them grow. By taking up nutrients, the plants purify the water. The clean water is then returned to the fishpond.

Just as in a natural pond ecosystem, the plants and animals in an aquaponics system are connected by cycles of water and nitrogen.

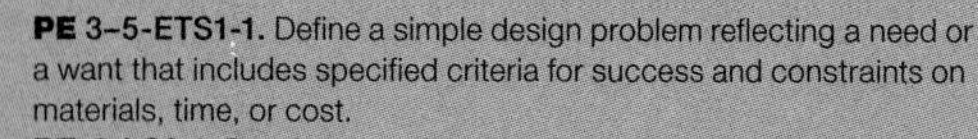

PE 3–5-ETS1-1. Define a simple design problem reflecting a need or a want that includes specified criteria for success and constraints on materials, time, or cost.
PE 5-LS2-1. Develop a model to describe the movement of matter among plants, animals, decomposers, and the environment.

This aquaponics system uses the wastewater from fishponds to fertilize crops.

The Challenge

Your engineering challenge is to design and build a model of an aquaponics system that could provide food for a community. Your system must:

- have two compartments—one for plants and one for aquatic organisms
- use a wick to move water to the plants
- keep the plants and the aquatic organisms alive for at least three weeks

STEM
ENGINEERING PROJECT
(continued)

1 Define the problem. 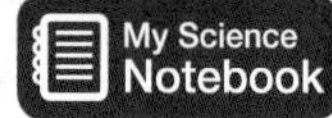

Think about the problem you are solving. What does your aquaponics system need to do? Look at the challenge box on the previous page. These are the criteria of your problem. Criteria tell you if your design is successful.

There are also constraints, or limits to your design. Your teacher will show you the materials you can use to build your system. You cannot use any other materials.

Your teacher will give you a container made from a two-liter plastic bottle. Your system must fit inside this container, but your plants may grow taller. During the test, you may add water to the bottom of your container, but you cannot add water to the top.

Write the problem you need to solve in your science notebook. Also list the criteria and constraints of the problem.

2 Find a solution.

Observe the container for your system. Your plants will grow in the compartment on top. The aquatic organisms will live at the bottom.

Look at the other materials. Which materials could be used to support the roots of the plants? Which materials could hold that substrate in place?

Pick three materials that could wick water up to the plants. Conduct fair tests to determine which of them would make the best wick. Record your results in your science notebook.

Think of a design. Sketch your plan, and present it to your team. Discuss each plan. Choose the design that will best meet the criteria and constraints of the problem.

Draw your team's final design. Label the materials, and explain why you choose them. Have your teacher approve your plan.

Fish in the pond release wastes into the water. Microbes and worms convert the wastes to nitrogen that plants can use. Plants take up the nitrogen and filter the water. Clean water is returned to the fishpond.

Some small aquaponic systems are designed for home use.

3 Test your solution.

Follow your plan to build the upper compartment of your aquaponics system. Insert the wick.

Pour 500 milliliters of clean water into the bottom of your container. Place the aquatic organisms in that water. Insert the upper compartment into the container. Use some clean water to moisten your substrate.

Your teacher will give you several young plants. Poke holes in the substrate, and gently place the roots of the plants in the holes. Measure the height of the plants, and count their leaves. Record your observations.

Place your aquaponics system in a well-lit location. Check it regularly. Add water to the bottom as needed.

After three weeks, observe your system. Are the plants and aquatic organisms still healthy? Measure the height of the plants, and count their leaves. Record your observations.

4 Refine your solution.

Present your aquaponics system to the class. Identify the materials you used. Did your system meet the needs of your plants and aquatic organisms? Use evidence to support your claim.

Compare your system with those made by other teams. Make a class chart comparing the height of the plants and the number of leaves on them. Which team's plants grew the most?

With your team, discuss the success of your design. Also talk about the other teams' designs. Identify at least two ways you could change your aquaponics system to make it more successful.

Make a poster showing the design of your aquaponics system. List the changes you would like to make. Explain how those changes would improve your design. Your poster should also use arrows to show the movement of matter through your system.

Plants Invade!

Sometimes a newly introduced species can damage the balance of an ecosystem. Kudzu is a plant that is native to Japan and southern China. People brought kudzu to the United States in the late 1800s. Gardeners planted it because of its pretty vines and sweet smell. Farmers planted it to help control erosion.

Kudzu was nicknamed "the vine that ate the South."

DCI LS2.A: Interdependent Relationships in Ecosystems. The food of almost any kind of animal can be traced back to plants. Organisms are related in food webs in which some animals eat plants for food and other animals eat the animals that eat plants. Some organisms, such as fungi and bacteria, break down dead organisms (both plants or plants parts and animals) and therefore operate as "decomposers." Decomposition eventually restores (recycles) some materials back to the soil. Organisms can survive only in environments in which their particular needs are met. A healthy ecosystem is one in which multiple species of different types are each able to meet their needs in a relatively stable web of life. Newly introduced species can damage the balance of an ecosystem. (5-LS2-1)

CCC Systems and System Models. A system can be described in terms of its components and their interactions. (5-LS2-1), (5-ESS2-1), (5-ESS3-1)

But after kudzu was introduced, it began to crowd out native trees and shrubs. It has now spread throughout much of the southeastern United States. Kudzu is an invasive species. An **invasive species** is an organism that does not belong in a certain place and harms the environment. Kudzu grows quickly, covering other plants, utility poles and power lines, and even buildings.

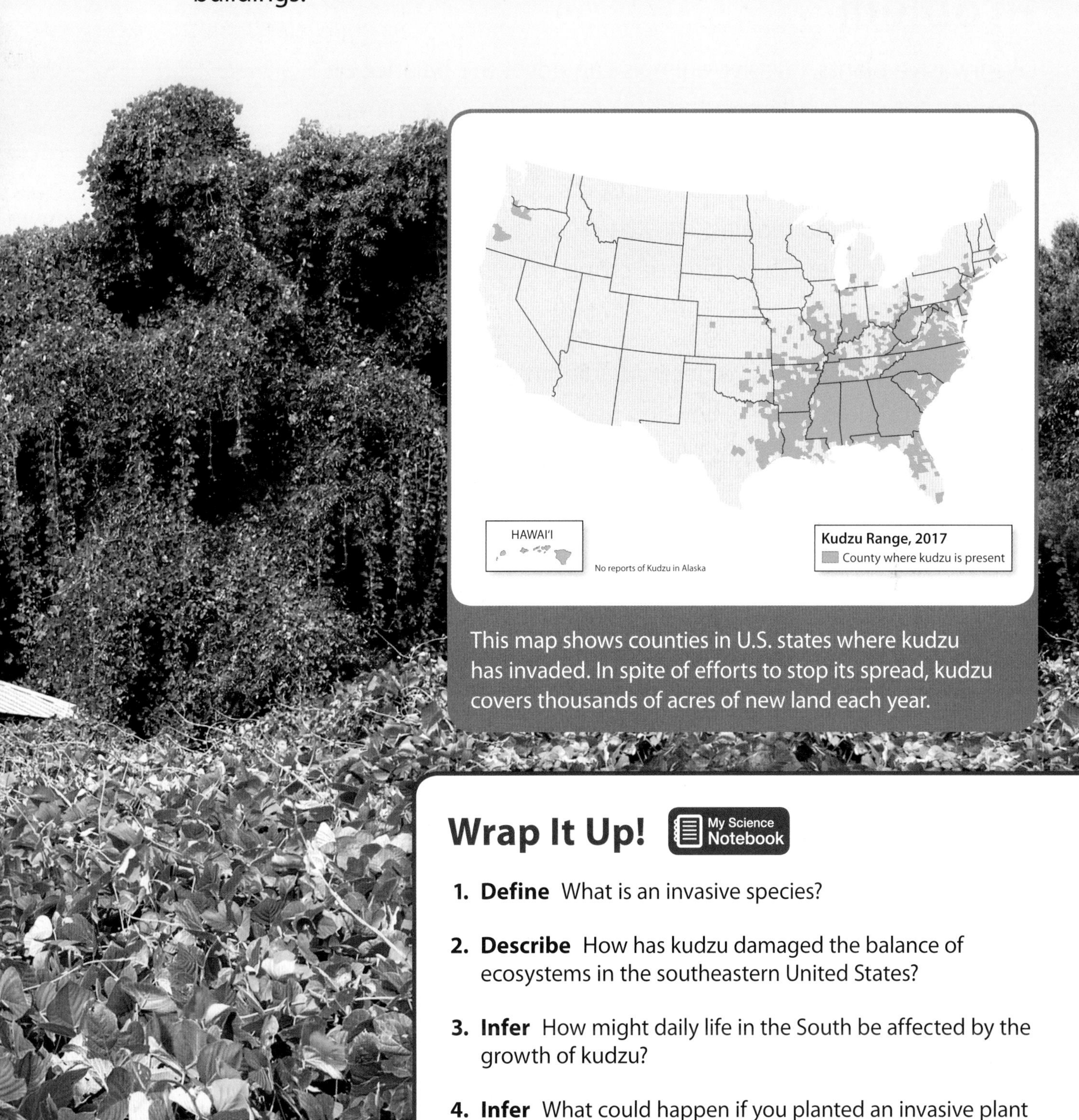

This map shows counties in U.S. states where kudzu has invaded. In spite of efforts to stop its spread, kudzu covers thousands of acres of new land each year.

Wrap It Up! My Science Notebook

1. **Define** What is an invasive species?
2. **Describe** How has kudzu damaged the balance of ecosystems in the southeastern United States?
3. **Infer** How might daily life in the South be affected by the growth of kudzu?
4. **Infer** What could happen if you planted an invasive plant near your home?

NATIONAL GEOGRAPHIC | **Think Like a Scientist**
Case Study

Animals Invade!

The red imported fire ant is native to South America.

Problem

Like invasive plants, invasive animals can upset the balance of an ecosystem. The red imported fire ant is an invasive animal that is causing problems in the United States. These ants have a painful sting. They can kill native insects, spiders, young birds, mammals, and reptiles. They can reduce the population of native ant species. They damage crops and can harm or kill young farm animals.

Red imported fire ants have been invading more and more areas of the United States. They don't have any natural enemies in the United States. How can this invasive species be controlled?

DCI LS2.A: Interdependent Relationships in Ecosystems. The food of almost any kind of animal can be traced back to plants. Organisms are related in food webs in which some animals eat plants for food and other animals eat the animals that eat plants. Some organisms, such as fungi and bacteria, break down dead organisms (both plants or plants parts and animals) and therefore operate as "decomposers." Decomposition eventually restores (recycles) some materials back to the soil. Organisms can survive only in environments in which their particular needs are met. A healthy ecosystem is one in which multiple species of different types are each able to meet their needs in a relatively stable web of life. Newly introduced species can damage the balance of an ecosystem. (5-LS2-1)
CCC Systems and System Models. A system can be described in terms of its components and their interactions. (5-LS2-1), (5-ESS2-1), (5-ESS3-1)

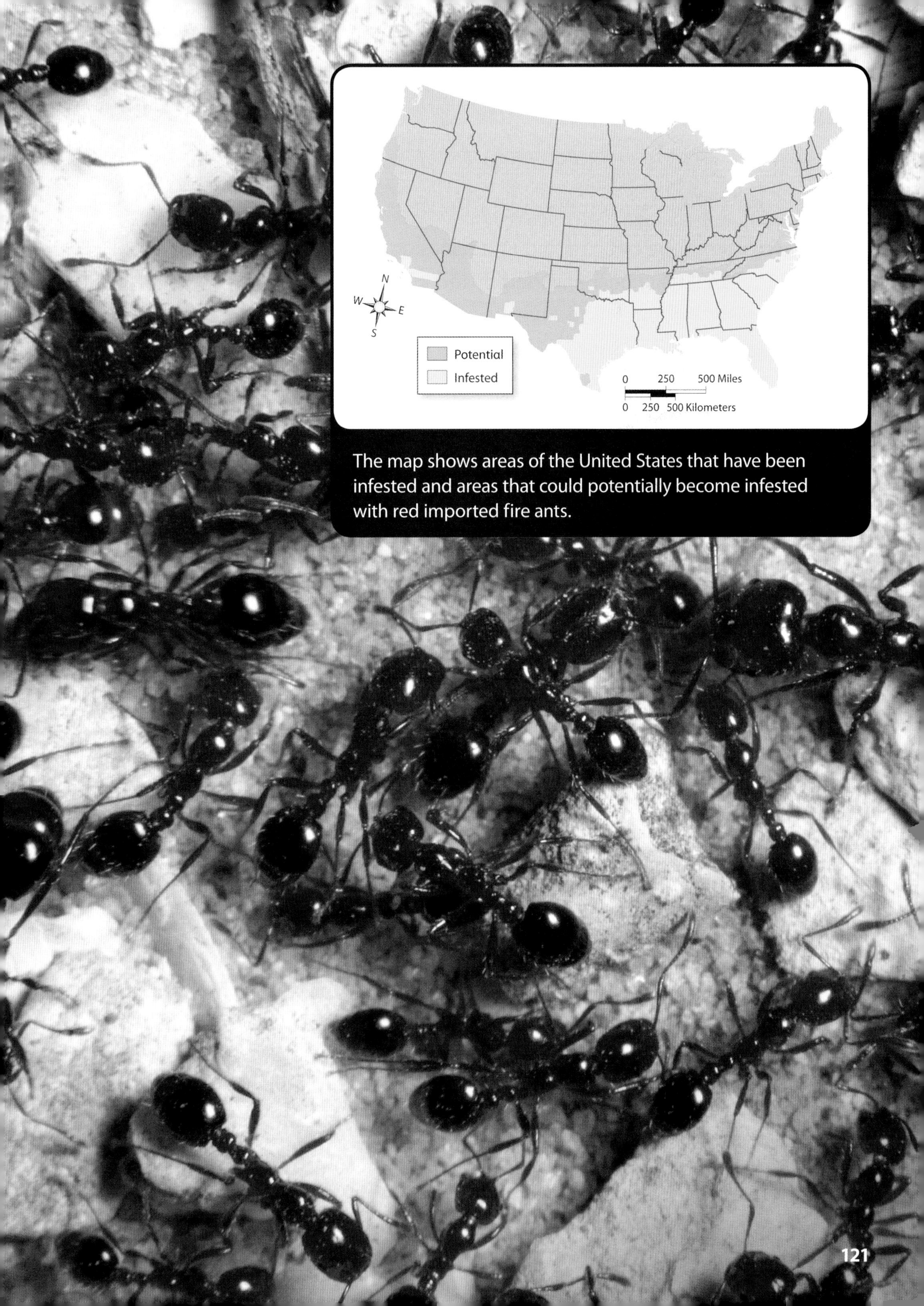

The map shows areas of the United States that have been infested and areas that could potentially become infested with red imported fire ants.

Solution

Recently scientists have been raising and releasing phorid flies in regions that have been invaded by red imported fire ants. Phorid flies are from South America, where they are a natural enemy of red fire ants.

phorid fly

When phorid flies come near red imported fire ants, the ants try to hide. The flies keep the ants from searching for food. With less food, the ant population cannot grow. But why do the ants hide from phorid flies? It seems that around phorid flies, red imported fire ants just can't keep their heads on straight! Read on to find out why.

1 A phorid fly lays an egg in the body of an ant. The egg hatches and the larva moves into the ant's head. It begins to eat the brain. Like a zombie, the ant wanders away from the nest.

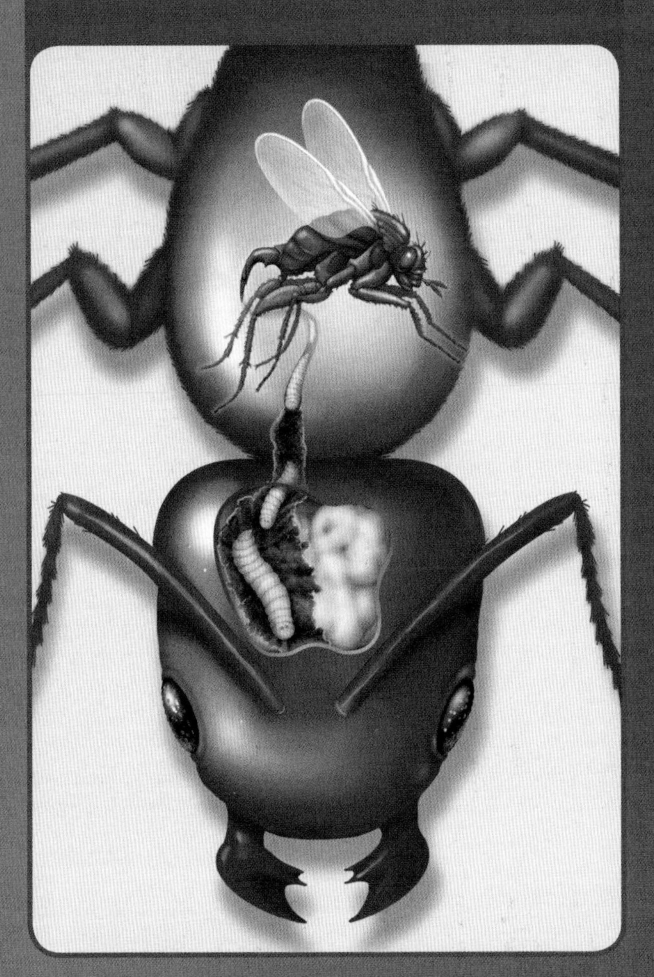

2 The larva releases a chemical that causes the ant's head to fall off. The larva continues to feed and develops into a pupa.

3 Finally the adult phorid fly emerges from the ant's head.

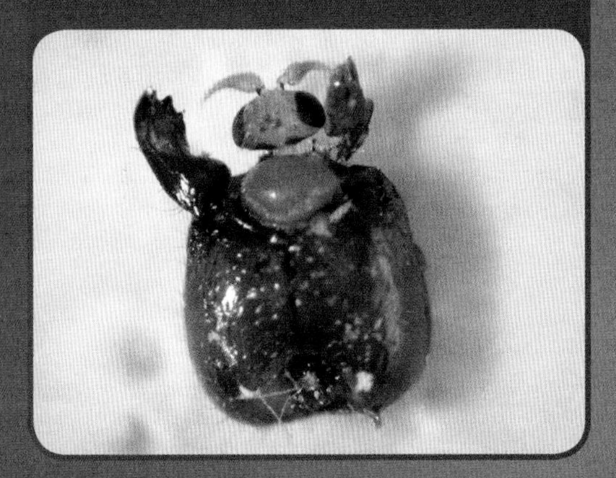

Wrap It Up!

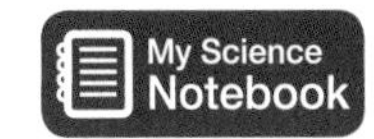

1. **Recall** What is an invasive species?

2. **Describe** How have red imported fire ants damaged the balance of ecosystems in the southeastern United States?

3. **Cause and Effect** How do phorid flies affect the behavior of red imported fire ants? How does this affect the red imported fire ant population?

4. **Evaluate** Are phorid flies an invasive species? Explain.

To the Treetops

An injury left Rebecca paralyzed from the waist down. Through her own strength and determination, she climbs high in the trees.

Rebecca and other researchers collected specimens from the canopy. They shared their discoveries with other scientists.

DCI LS2.A: Interdependent Relationships in Ecosystems. The food of almost any kind of animal can be traced back to plants. Organisms are related in food webs in which some animals eat plants for food and other animals eat the animals that eat plants. Some organisms, such as fungi and bacteria, break down dead organisms (both plants or plants parts and animals) and therefore operate as "decomposers." Decomposition eventually restores (recycles) some materials back to the soil. Organisms can survive only in environments in which their particular needs are met. A healthy ecosystem is one in which multiple species of different types are each able to meet their needs in a relatively stable web of life. Newly introduced species can damage the balance of an ecosystem. (5-LS2-1)

NS Science Models, Laws, Mechanisms, and Theories Explain Natural Phenomena. Science explanations describe the mechanisms for natural events. (5-LS2-1)

"If you've got something that you're passionate about, . . . there's always a way to do what you love and what you care about." —Rebecca Tripp

Rebecca Tripp recently traveled to the Amazon rain forest. She collected data on the relationship between leaves and the insects that eat them. She does her work in the canopy. The canopy is the top layer of a forest. This is the layer of leaves and branches of the tallest trees. Using ropes and a harness to reach the highest treetops is something that Rebecca never imagined she would be doing. Although she has always been interested in nature, an injury left her paralyzed from the waist down. She is dependent on a wheelchair. For a long time, she thought being in a wheelchair would stop her from achieving her dreams.

Rebecca grew up in Maine, developing her love of nature as a young girl. As she grew older, she wanted to work to conserve and preserve the natural world. After her injury, however, she was filled with doubt about pursuing a career in conservation. Then she found a science internship funded by the National Science Foundation that was actively looking for people with mobility disabilities. She quickly signed up for the summer that changed her life.

That summer, she learned how to use ropes and a harness to climb trees and reach the canopy. There, she collected samples of mosses, lichens, and leaves. She learned how to process the samples and analyze data. By the end of the summer, the team of interns and scientists had collected and identified over 4,000 samples. They possibly found four previously unknown species. The team continued their practice of science by sharing the results of their work with other scientists at the North Carolina Museum of Natural Sciences.

Rebecca received a bachelor's degree in psychology, but has returned to school to study biology. She is pursuing her original passion of conserving nature. She also has reported her personal story in several essays. Rebecca hopes that sharing her story will raise awareness that it is possible for people with disabilities to work in field biology or other areas of science. She hopes to inspire others to never give up on their dreams. In an interview during her trip to the Amazon she said, "Anything is possible, whether you are disabled or able-bodied. . . . If you've got something that you're passionate about, there is always a way to do it. . . . There's always a way to do what you love and what you care about."

Wrap It Up!

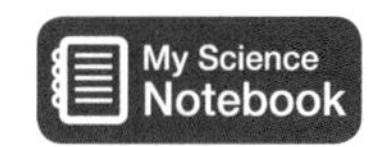

1. **Explain** What methods of science has Rebecca used to complete her work?
2. **Relate** Explain a time when you have been filled with doubt and what you did to overcome it.
3. **Infer** Why do you think it is important for Rebecca to tell her story to others?

Conservationist

The Colorado River is the largest river in the Southwest. It flows through many states before reaching its mouth, or **delta,** in Mexico. Along the way, people use its water for farms, lawns, cities, and factories. Over the last 100 years, people have been taking more and more water from the river. Now people take so much water that by the time the Colorado River reaches the sea, it is only a trickle. There isn't enough water left for the wildlife that live in the delta.

Dr. Osvel Hinojosa is a conservationist who is working with many other people to restore the river's flow. He is also working to restore the wildlife of the delta. For example, he has helped replace invasive saltcedar trees with native trees, such as willows, cottonwood, and mesquite.

NGL Science Why are the wetlands of the delta important?

Dr. Hinojosa Wetlands have a very important function in the world. Many species of plants and animals live in wetlands. Wetlands also help make the water cleaner and provide protection against floods, storms, and hurricanes.

NS Science Models, Laws, Mechanisms, and Theories Explain Natural Phenomena. Science explanations describe the mechanisms for natural events. (5-LS2-1)
NS Science Addresses Questions About the Natural and Material World. Science findings are limited to what can be answered with empirical evidence.

Osvel Hinojosa is a conservationist and National Geographic Explorer. Osvel works with businesses, environmental organizations, and governments in the United States and Mexico to restore the wetlands and other ecosystems of the Colorado River Delta.

This channel in the Colorado River Delta is nearly dried up. It cannot support the wetland wildlife typically found in a river delta.

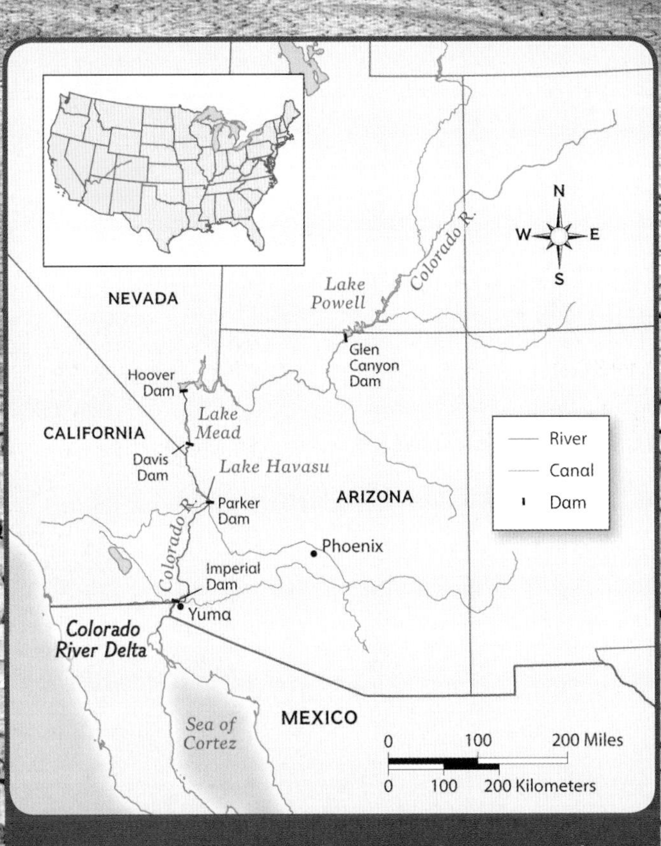

Many dams and canals cause changes in the flow of water in the Colorado River Delta.

NGL Science What is the problem that you are working to solve?

Dr. Hinojosa The main challenge we have is nature does not equally divide water resources. We have failed to recognize that all things in nature need water. The water needs to be available for more than just our agricultural industry. We also need to make sure that there is enough water flowing in the rivers to connect the rivers to the seas.

NGL Science Is it hard to work with all the different groups involved in restoring the river?

Dr. Hinojosa It takes time. But once you find common ground and make it clear that everyone is working toward a common goal, which is to improve conditions for everyone, then it's easier to make progress.

NGL Science How can students help save rivers and wetlands?

Dr. Hinojosa Learn about your watershed, where the water is produced, where the system goes, what are the important environmental values of your watershed and what are the conservation concerns of that watershed. Go out and get engaged with the groups that are out there doing great work. You can make a difference.

Volunteers plant cottonwoods and willows in a restoration site in the Colorado River Delta.

These ring-billed gulls are stopping over in the largest remaining wetland in the Colorado River Delta. This area is a critical wintering and stopover site for birds that migrate along the Pacific Flyway.

Check In My Science Notebook

Congratulations! You have completed *Life Science*. Now let's reflect on what you have learned. Here you'll find a checklist and some additional questions to help you assess your progress.

Page through your science notebook to find examples of different items. On a separate page in your science notebook, write your assessment of your work so far.

Read each item in this list. Ask yourself if you think you did a good job of it.

For each item, select the choice that is true for you: A. Yes **B.** Not Yet

- I defined and illustrated science vocabulary and main ideas.
- I labeled drawings. I wrote notes to explain ideas.
- I collected photos, news stories, and other objects.
- I used tables, charts, or graphs to record and analyze data.
- I included evidence for explanations and conclusions.
- I described how scientists and engineers answer questions and solve problems.
- I asked new questions of my own.
- I did something else. (Describe it.)

Reflect on Your Learning

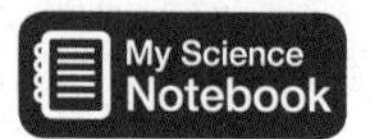

1. What did you learn that changed the way you think about something?
2. Choose one main science idea that you think was most important to learn about. Why was it important to you?

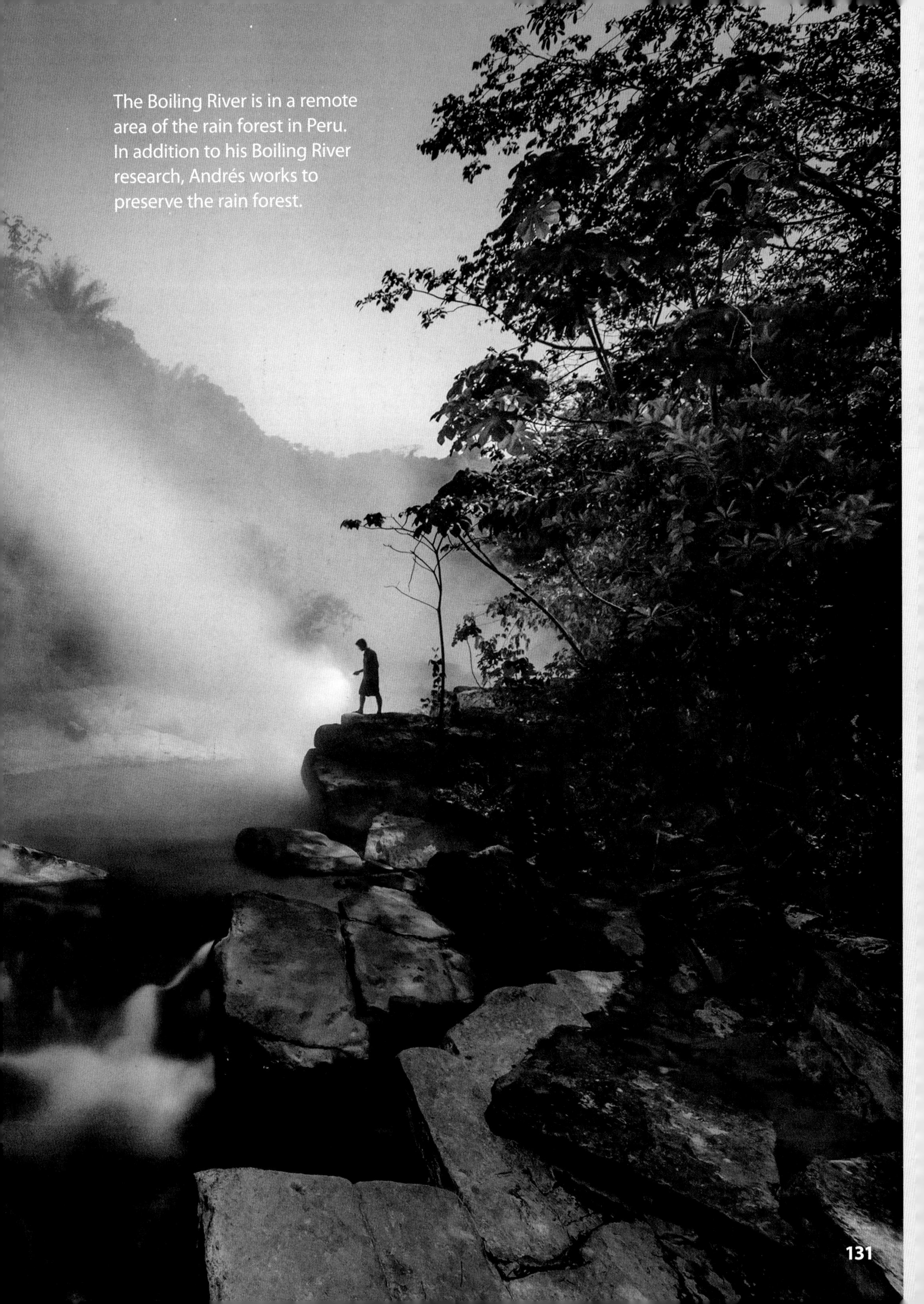

The Boiling River is in a remote area of the rain forest in Peru. In addition to his Boiling River research, Andrés works to preserve the rain forest.

Explorer

Andrés Ruzo Geoscientist
National Geographic Explorer

Let's Explore!

Science involves human creativity. Often, science brings together multiple disciplines—from art to math. Science is all about creativity and evidence! Your notebook is the "home" of your observations. If you have done a good job with your notebook, a student 100 years from now will be able to copy what you did exactly and will get the same result, assuming the thing you are observing has not changed.

In Earth science, I study volcanos and heat beneath Earth's surface. Some other areas of Earth science include weather and climate, landforms, natural resources, and even space. Here are some questions you might answer as you study *Earth Science*:

- What is the most common gas in the atmosphere that you breathe in every day?
- In one day, how much rain might a summer monsoon produce?
- What are tufa towers, and how do they form?
- How do wastes from homes, farms, mines, and factories affect water systems on Earth?
- How can scientists and engineers grow more trees in cities where space is scarce?
- What is space junk, and what problems might it cause?

Look at the drawings for some notebook examples and ideas. Let's check in again later to review what you have learned!

In your science notebook, define and illustrate main ideas.

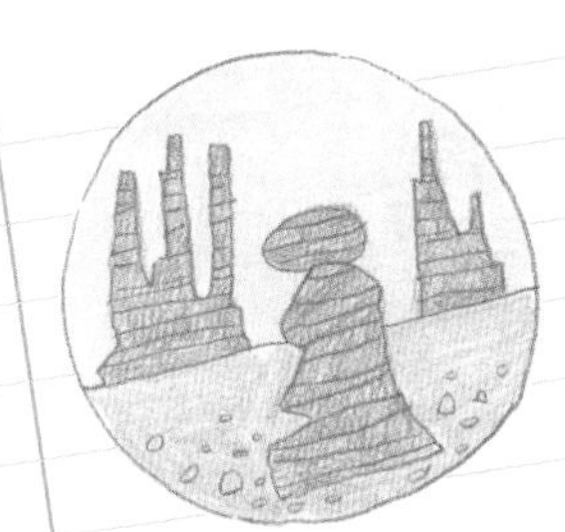

Geosphere – the solid outer part of Earth

Atmosphere – the layer of air around Earth

Hydrosphere – all the water near Earth's surface and the water vapor in the air

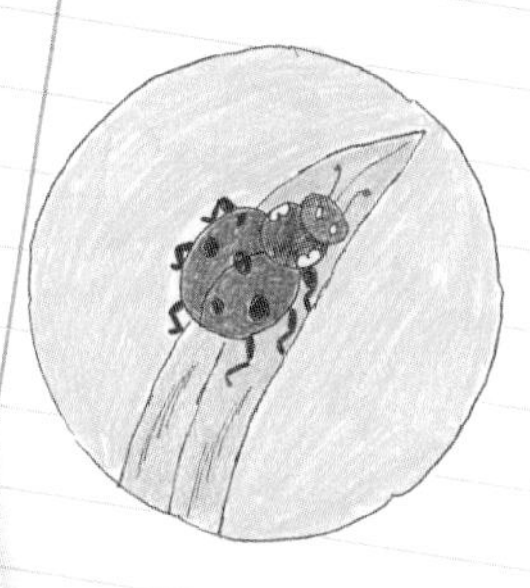

Biosphere – all the parts of Earth where life exists

Write about how what you learned changed your thinking about some science concepts.

I thought that the phases of the moon were caused by a shadow. Now I understand that the phases are caused by the positions of Earth, the moon, and the sun.

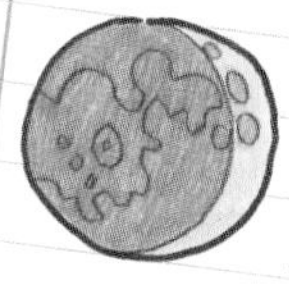

waxing crescent moon

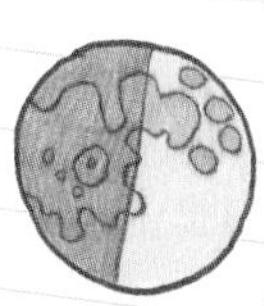

first quarter moon

full moon

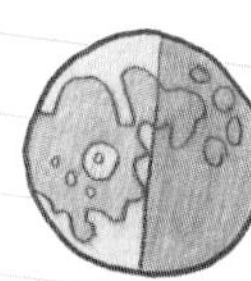

third quarter moon

waning crescent moon

Earth Science

Earth's Systems

Space Systems: Stars and the Solar System

This rock arch is a product of the interaction of Earth's systems.

Earth's Major Systems

Scientists who study Earth divide it into four major systems. These systems, or spheres, are the geosphere, hydrosphere, atmosphere, and biosphere. Each of these systems is made up of many parts that work

Geosphere

The **geosphere** contains all of Earth's solid and molten, or liquid, rocks. It also includes sediments and soil.

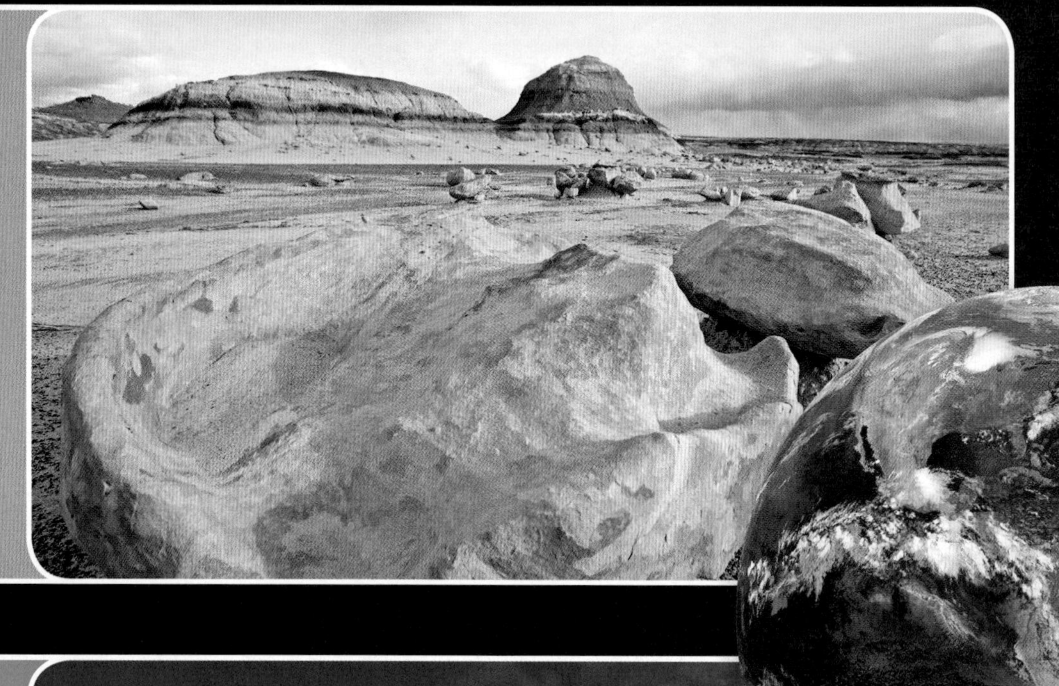

Hydrosphere

The **hydrosphere** is made up of all of the liquid water on Earth, such as the water in rivers, lakes, and the ocean, as well as water found underground. It also includes all of the ice and snow on Earth.

together as a whole. The four systems interact in many different ways. Their interactions are constantly changing Earth's surface materials and processes. Study the photos to learn more about these four systems.

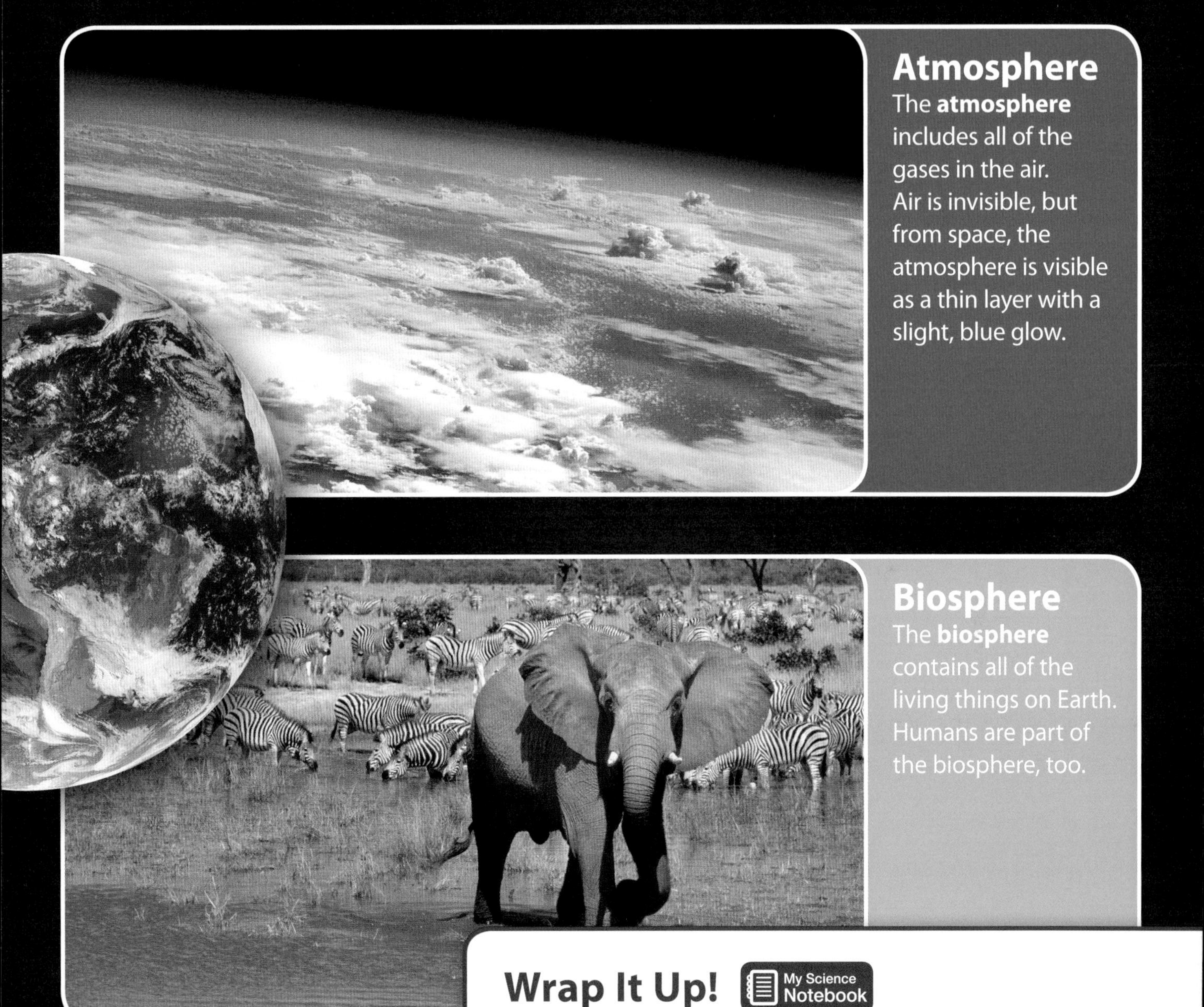

Atmosphere

The **atmosphere** includes all of the gases in the air. Air is invisible, but from space, the atmosphere is visible as a thin layer with a slight, blue glow.

Biosphere

The **biosphere** contains all of the living things on Earth. Humans are part of the biosphere, too.

Wrap It Up! My Science Notebook

1. **List** What are Earth's four major systems?
2. **Classify** Identify the system to which each of the following belongs: water in a pool, pebbles on a beach, a bee visiting a flower, the oxygen that you breathe.
3. **Apply** Beavers often build dams on streams, forming ponds. Which systems are interacting when a beaver builds a dam?

The Geosphere

The geosphere is made up of all of the rocks and minerals on Earth's surface. It includes all of the materials in mountain ranges, canyons, beaches, and other landforms. Sediments and soil are a part of the geosphere. The solid rocks and molten materials beneath the surface are also a part of the geosphere.

The geosphere is made up of all of the solid and liquid rock on Earth.

Volcanoes release lava, ash, and gases. These materials change the shape of the land and the composition of the atmosphere.

DCI ESS2.A: Earth Materials and Systems. Earth's major systems are the geosphere (solid and molten rock, soil, and sediments), the hydrosphere (water and ice), the atmosphere (air), and the biosphere (living things, including humans). These systems interact in multiple ways to affect Earth's surface materials and processes. The ocean supports a variety of ecosystems and organisms, shapes landforms, and influences climate. Winds and clouds in the atmosphere interact with the landforms to determine patterns of weather. (5-ESS2-1)
CCC Systems and System Models. A system can be described in terms of its components and their interactions. (5-ESS2-1)

Processes and events in the geosphere are constantly changing the shape of the land. For example, earthquakes crack and move the land. The geosphere also interacts with other systems. Over time, flowing water of the hydrosphere weathers rocks into small pieces. The water erodes the pieces and deposits them in other places. Plants in the biosphere grow in the sediments and soil.

Sediment is made as wind, water, and ice break or weather larger rocks into smaller pieces.

Wrap It Up! My Science Notebook

1. **Identify** What are two events or processes that take place in the geosphere?

2. **Cause and Effect** How can water in the hydrosphere change the geosphere?

3. **Explain** A volcano releases large amounts of ash into the air. How might this process affect plants and animals in the biosphere?

The Hydrosphere

GEOSPHERE ATMOSPHERE HYDROSPHERE BIOSPHERE

The hydrosphere is made up of all of the liquid water on Earth, as well as all of the water frozen in ice and snow. It includes the fresh water in streams, rivers, ponds, lakes, and wetlands. It includes the salt water in the ocean. The clouds in the sky, which are made up of water droplets or particles of ice, are also part of the hydrosphere. **Groundwater**—the water in soil and between the rocks below Earth's surface—is a part of the hydrosphere, too.

The water in the hydrosphere is constantly moving. Rain that falls on land enters rivers, which flow toward the ocean. Water enters the atmosphere when it **evaporates**, or changes from a liquid to a gas. Eventually water

Clouds Clouds are made up of tiny water droplets or ice crystals.

DCI **ESS2.A: Earth Materials and Systems.** Earth's major systems are the geosphere (solid and molten rock, soil, and sediments), the hydrosphere (water and ice), the atmosphere (air), and the biosphere (living things, including humans). These systems interact in multiple ways to affect Earth's surface materials and processes. The ocean supports a variety of ecosystems and organisms, shapes landforms, and influences climate. Winds and clouds in the atmosphere interact with the landforms to determine patterns of weather. (5-ESS2-1)

CCC **Systems and System Models.** A system can be described in terms of its components and their interactions. (5-ESS2-1)

in the atmosphere **condenses** onto dust and other tiny particles in the air, forming clouds. Liquid or solid water falls to the ground as rain or snow.

All organisms require water to survive. A great variety of organisms live in the ocean, while others live in fresh water. On land, plants take in water from the soil. All animals need water to survive. Most land animals cannot survive without taking in water.

Icebergs Icebergs are made up of the solid state of water—ice! Similar to the way ice floats in your lemonade, icebergs float in ocean water.

Salt water The salt water in the ocean is part of the hydrosphere. Most of the water on Earth is found in the ocean. The ocean supports a variety of ecosystems.

Wrap It Up!

1. **Define** What is groundwater?
2. **Identify** What process moves water from Earth's surface to the atmosphere?
3. **Explain** How does water in the atmosphere return to Earth's surface?

The Atmosphere

The atmosphere is the thin layer of gases that surround Earth. The circle graph shows the relative amounts of the major atmospheric gases. The atmosphere also contains water vapor, or water that has evaporated and is a gas.

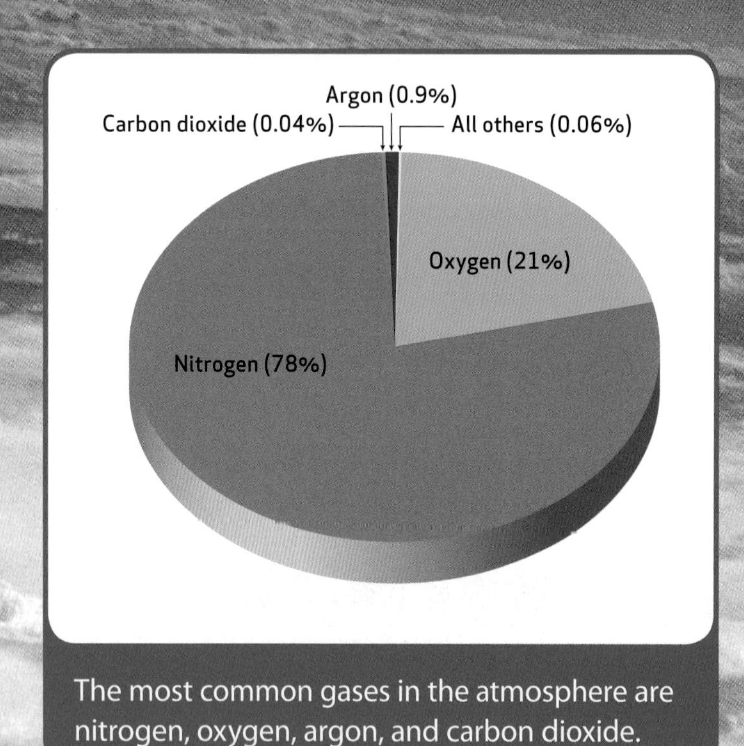

The most common gases in the atmosphere are nitrogen, oxygen, argon, and carbon dioxide.

DCI **ESS2.A: Earth Materials and Systems.** Earth's major systems are the geosphere (solid and molten rock, soil, and sediments), the hydrosphere (water and ice), the atmosphere (air), and the biosphere (living things, including humans). These systems interact in multiple ways to affect Earth's surface materials and processes. The ocean supports a variety of ecosystems and organisms, shapes landforms, and influences climate. Winds and clouds in the atmosphere interact with the landforms to determine patterns of weather. (5-ESS2-1)

CCC **Systems and System Models.** A system can be described in terms of its components and their interactions. (5-ESS2-1)

The gases in the atmosphere capture some of the energy of sunlight. They also trap heat given off by Earth's surface. This keeps the planet warm. Certain gases protect living things from the sun's harmful rays.

The processes of weather, such as winds and storms, take place in the atmosphere. Winds move sediments across Earth's surface, forming sand dunes and other landforms. Winds blow across lakes and the ocean, forming waves that crash onto the land. Storms produce rain and snow, filling rivers and lakes and changing the surface of the land.

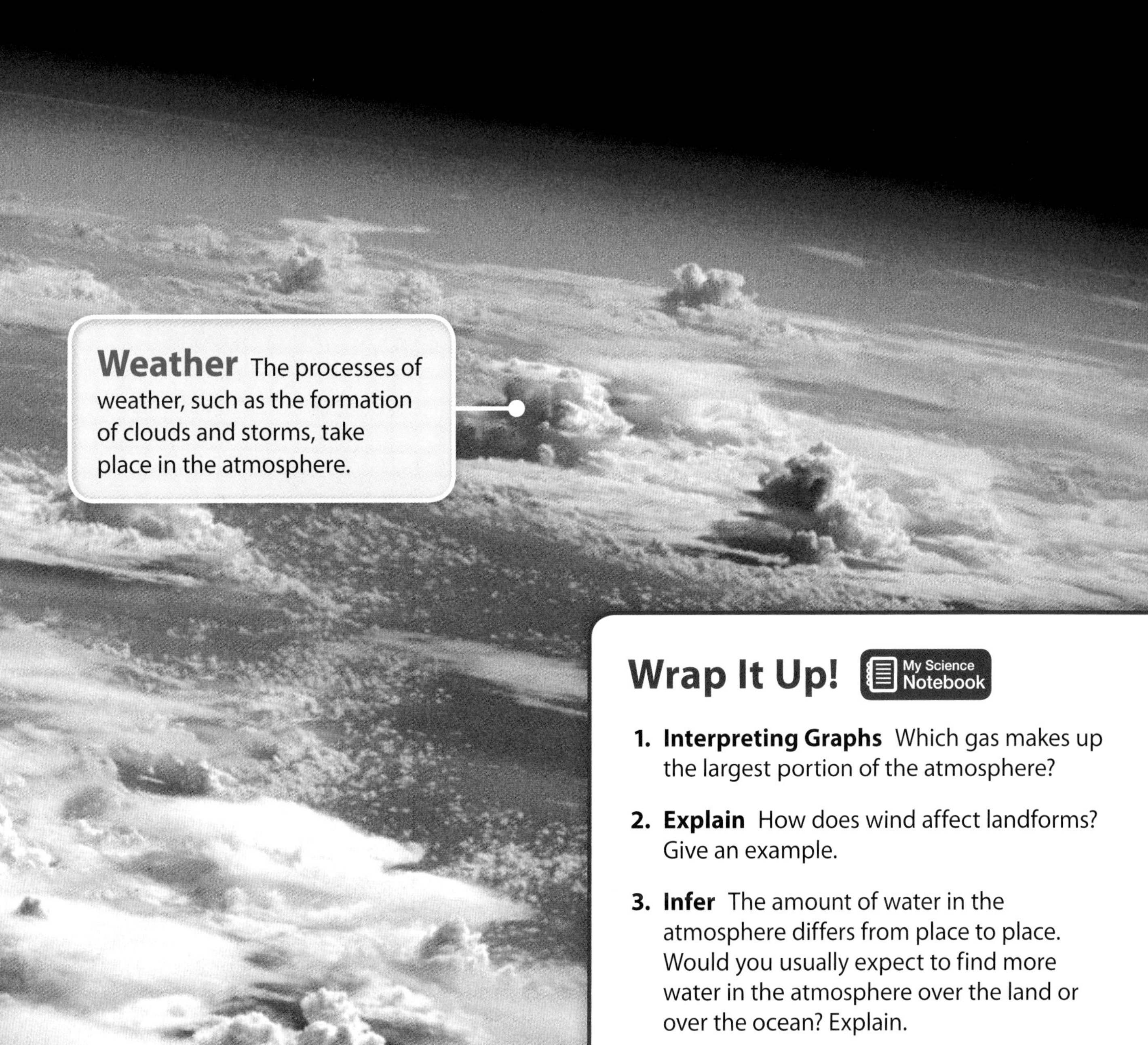

Weather The processes of weather, such as the formation of clouds and storms, take place in the atmosphere.

Wrap It Up! My Science Notebook

1. **Interpreting Graphs** Which gas makes up the largest portion of the atmosphere?
2. **Explain** How does wind affect landforms? Give an example.
3. **Infer** The amount of water in the atmosphere differs from place to place. Would you usually expect to find more water in the atmosphere over the land or over the ocean? Explain.

The Biosphere

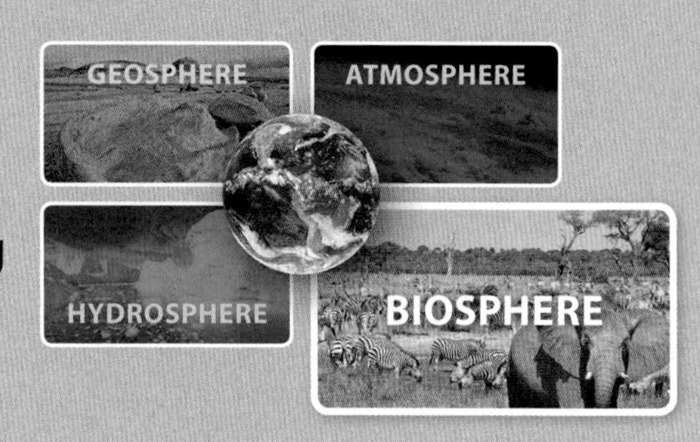

The biosphere is made up of all of the organisms living on Earth. It includes all of the plants, animals, fungi, and microbes living. The biosphere also includes human beings.

All living things require water and nutrients to live. Organisms that live in streams, lakes, or the ocean get water from their surroundings. On land, plants get water and nutrients from the soil. Most land animals get water by drinking.

All organisms need water from the hydrosphere.

DCI ESS2.A: Earth Materials and Systems. Earth's major systems are the geosphere (solid and molten rock, soil, and sediments), the hydrosphere (water and ice), the atmosphere (air), and the biosphere (living things, including humans). These systems interact in multiple ways to affect Earth's surface materials and processes. The ocean supports a variety of ecosystems and organisms, shapes landforms, and influences climate. Winds and clouds in the atmosphere interact with the landforms to determine patterns of weather. (5-ESS2-1)

CCC Systems and System Models. A system can be described in terms of its components and their interactions. (5-ESS2-1)

Plants take in carbon dioxide from the atmosphere to make food. They release oxygen into the atmosphere. Animals and plants use oxygen and release carbon dioxide when they break down food to release energy.

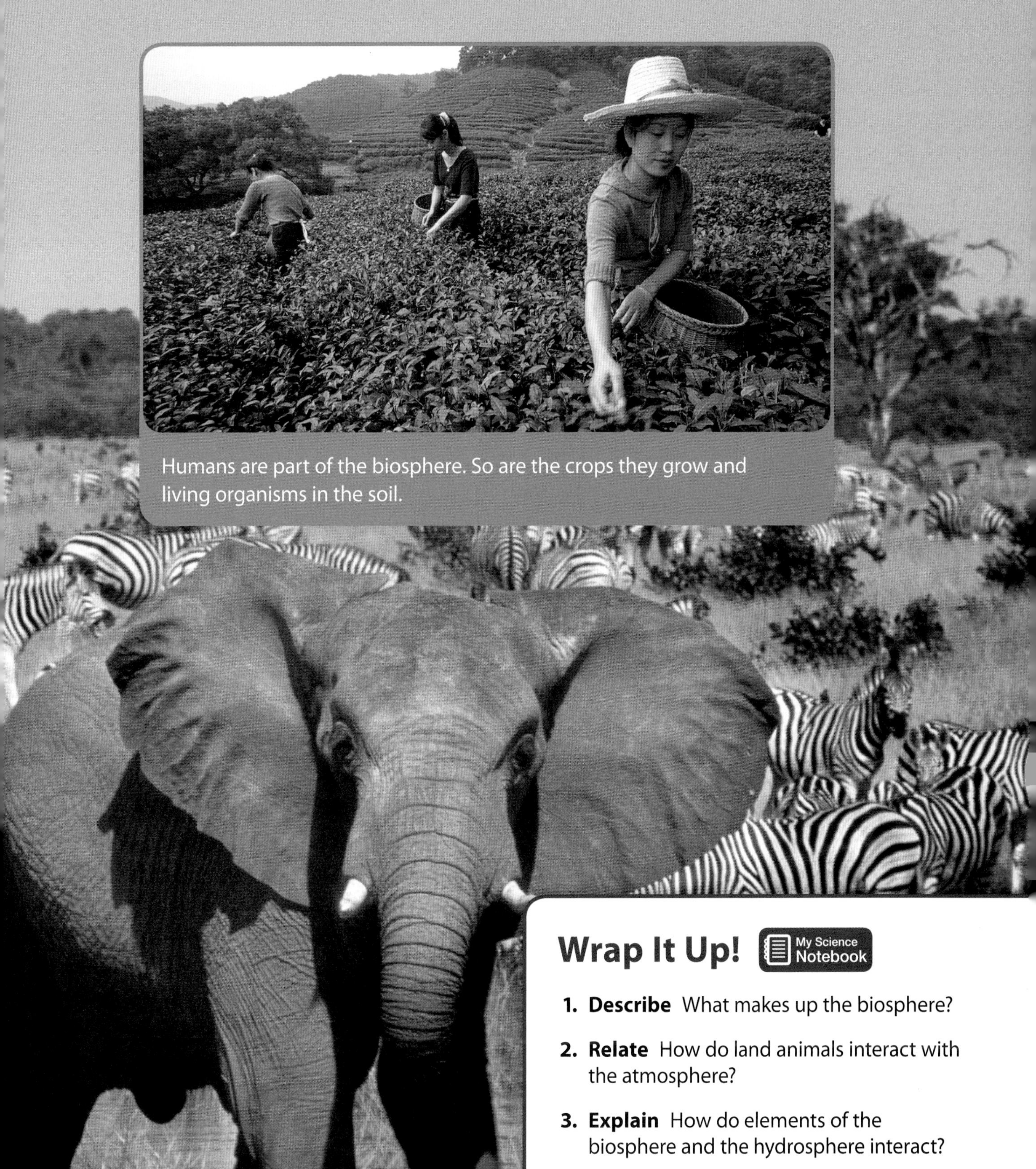

Humans are part of the biosphere. So are the crops they grow and living organisms in the soil.

Wrap It Up! My Science Notebook

1. **Describe** What makes up the biosphere?
2. **Relate** How do land animals interact with the atmosphere?
3. **Explain** How do elements of the biosphere and the hydrosphere interact?

Earth's Systems Interact

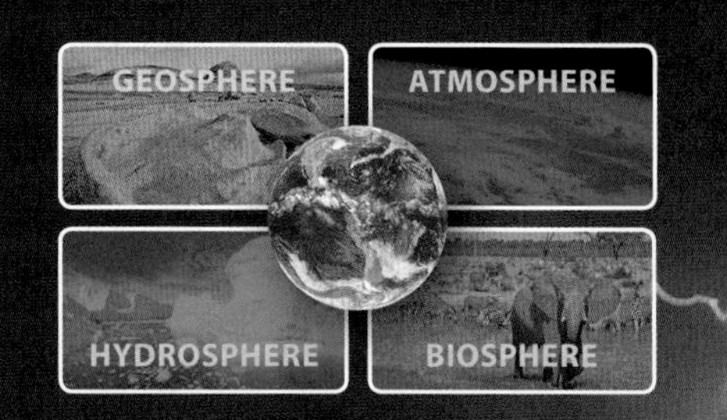

The monsoons of Southeast Asia and India have a powerful effect on the land, ecosystems, and people of the region. A **monsoon** is a strong wind that changes direction with the seasons. These seasonal changes cause the wet and dry seasons of Asia.

Monsoons result from the unequal heating of the land and the ocean. In winter the land is colder than the water. Dry monsoons blow from the land toward the ocean. Plants wither and wells run dry. In summer the water is cooler than the land. The monsoons blow from the ocean toward the land. They bring humid air and large amounts of rain. The rain causes plants to grow rapidly. Humans rely on the summer monsoons to provide water for rice and other crops.

The heavy rains of the summer monsoons may cause damage. Rushing streams sweep away soil. Low-lying areas flood, destroying homes and crops. People and livestock are at risk of drowning.

DCI ESS2.A: Earth Materials and Systems. Earth's major systems are the geosphere (solid and molten rock, soil, and sediments), the hydrosphere (water and ice), the atmosphere (air), and the biosphere (living things, including humans). These systems interact in multiple ways to affect Earth's surface materials and processes. The ocean supports a variety of ecosystems and organisms, shapes landforms, and influences climate. Winds and clouds in the atmosphere interact with the landforms to determine patterns of weather. (5-ESS2-1)
CCC Systems and System Models. A system can be described in terms of its components and their interactions. (5-ESS2-1)

Summer monsoons can dump as much as 30 centimeters (about 1 foot) of rain in a day!

The heavy rains of the summer monsoons flood low-lying fields.

Water that fell during the summer monsoons permits people to grow crops, such as rice, during the drier winter season.

Wrap It Up!

1. **Define** What is a monsoon?
2. **Compare and Contrast** How are the winter and summer monsoons alike and different? What causes them? Which bring rain and which bring dry air?
3. **Cause and Effect** How do the summer monsoons of India affect the hydrosphere, geosphere, and biosphere?

Investigate

Interactions of Earth's Systems

? How can you model interaction among Earth's major systems?

Earth's four spheres, or systems, interact in multiple ways. You can make a terrarium to model how the geosphere, atmosphere, hydrosphere, and biosphere interact.

Materials

clear plastic bottle with top cut off

gravel

potting soil

plastic spoon

small plants

water

masking tape

DCI ESS2.A: Earth Materials and Systems. Earth's major systems are the geosphere (solid and molten rock, soil, and sediments), the hydrosphere (water and ice), the atmosphere (air), and the biosphere (living things, including humans). These systems interact in multiple ways to affect Earth's surface materials and processes. The ocean supports a variety of ecosystems and organisms, shapes landforms, and influences climate. Winds and clouds in the atmosphere interact with the landforms to determine patterns of weather. (5-ESS2-1)
CCC Systems and System Models. A system can be described in terms of its components and their interactions. (5-ESS2-1)

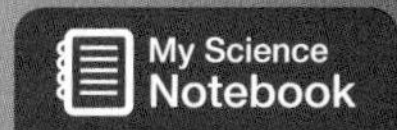

1 Spread about 2.5 cm of gravel in the bottom of the plastic bottle. Add 8–10 cm of potting soil.

2 Use the spoon to dig a hole for each plant. Gently place the plant's roots in the soil. Tap down the soil. Add water until it covers the gravel layer.

3 Carefully place the top of the bottle into the base of the bottle. Use masking tape to seal the terrarium. Make a labeled drawing of your terrarium in your notebook.

4 Place your terrarium in an area with a medium amount of light. Observe your terrarium once per week for two weeks or longer. Record your observations.

Wrap It Up!

1. **Making Models** Which materials in your terrarium represent each of Earth's four major systems?
2. **Explain** How did the plants in your terrarium interact with the hydrosphere?
3. **Infer** After a time, drops began to collect on the plastic bottle. What two processes in the atmosphere allowed water to collect there?

Ocean Ecosystems

Ocean waters cover more than 70 percent of Earth's surface. This vast area contains a great variety of environments, each of which supports a different ecosystem. From rocky shores to the deepest waters, the ocean is home to organisms of every color, shape, and size.

The humpback whale is larger than any animal that lives on land, but it is not the largest of the whales in the ocean.

DCI ESS2.A: Earth Materials and Systems. Earth's major systems are the geosphere (solid and molten rock, soil, and sediments), the hydrosphere (water and ice), the atmosphere (air), and the biosphere (living things, including humans). These systems interact in multiple ways to affect Earth's surface materials and processes. The ocean supports a variety of ecosystems and organisms, shapes landforms, and influences climate. Winds and clouds in the atmosphere interact with the landforms to determine patterns of weather. (5-ESS2-1)
CCC Systems and System Models. A system can be described in terms of its components and their interactions. (5-ESS2-1)

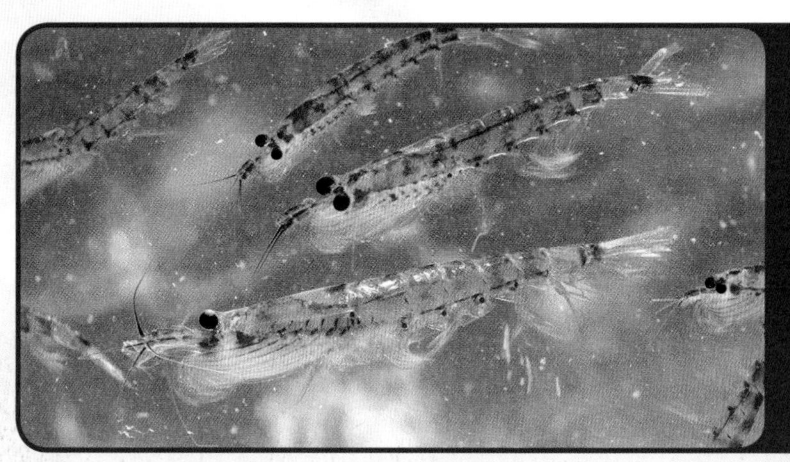

The open ocean is home to creatures of all sizes. These Antarctic krill are only 4–5 centimeters (about 2 inches) in length. Whales, seals, and penguins eat them.

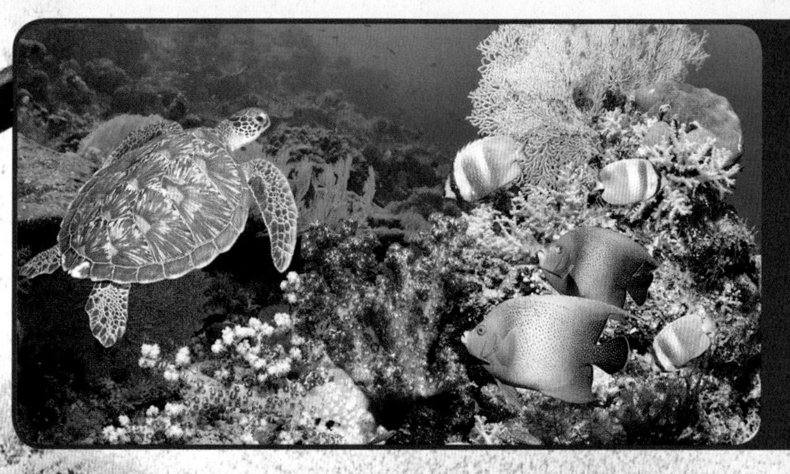

Coral reefs are home to small animals called corals and to the greatest variety of ocean life. Some corals produce limestone coverings that build the stony reefs in shallow tropical water.

No sunlight reaches the deep ocean. The deep sea anglerfish attracts prey there with a glowing lure on its head. The photographer lit up the darkness to take this picture.

Wrap It Up!

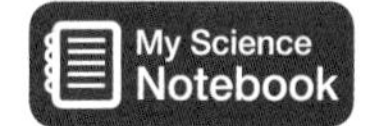

1. **Name** What are three ocean ecosystems?
2. **Compare** Which ecosystem supports a greater variety of living organisms, the deep ocean or a coral reef?
3. **Generalize** Why is the ocean able to support many different kinds of ecosystems?

The Ocean Shapes the Land

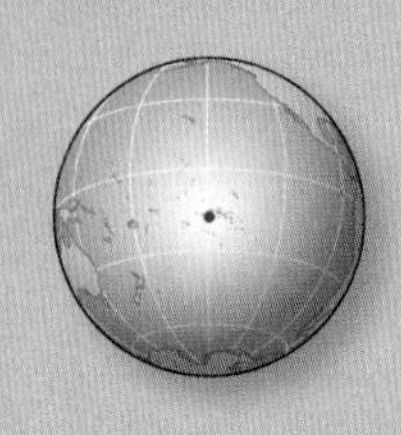

The ocean surf continues to shape these cliffs on the Oregon coast.

Waves and ocean currents are forces that shape landforms along the coast. **Ocean currents** are steady flows of ocean water. Moving ocean water helped shape the landforms you see in this photograph of Tahiti.

DCI **ESS2.A: Earth Materials and Systems.** Earth's major systems are the geosphere (solid and molten rock, soil, and sediments), the hydrosphere (water and ice), the atmosphere (air), and the biosphere (living things, including humans). These systems interact in multiple ways to affect Earth's surface materials and processes. The ocean supports a variety of ecosystems and organisms, shapes landforms, and influences climate. Winds and clouds in the atmosphere interact with the landforms to determine patterns of weather. (5-ESS2-1)
CCC **Systems and System Models.** A system can be described in terms of its components and their interactions. (5-ESS2-1)

In many coastal areas, waves crash onto the shore, wearing down cliffs and breaking rocks apart. The force of the waves may carve rocks into strange shapes. Waves and the sediments they carry eventually grind rocks into tiny bits of sand.

In the process of **erosion,** waves and currents pick up sediments and move them away. In the process of **deposition,** sand and other sediments are laid down in a new place. Deposition builds up on sandy islands and beaches.

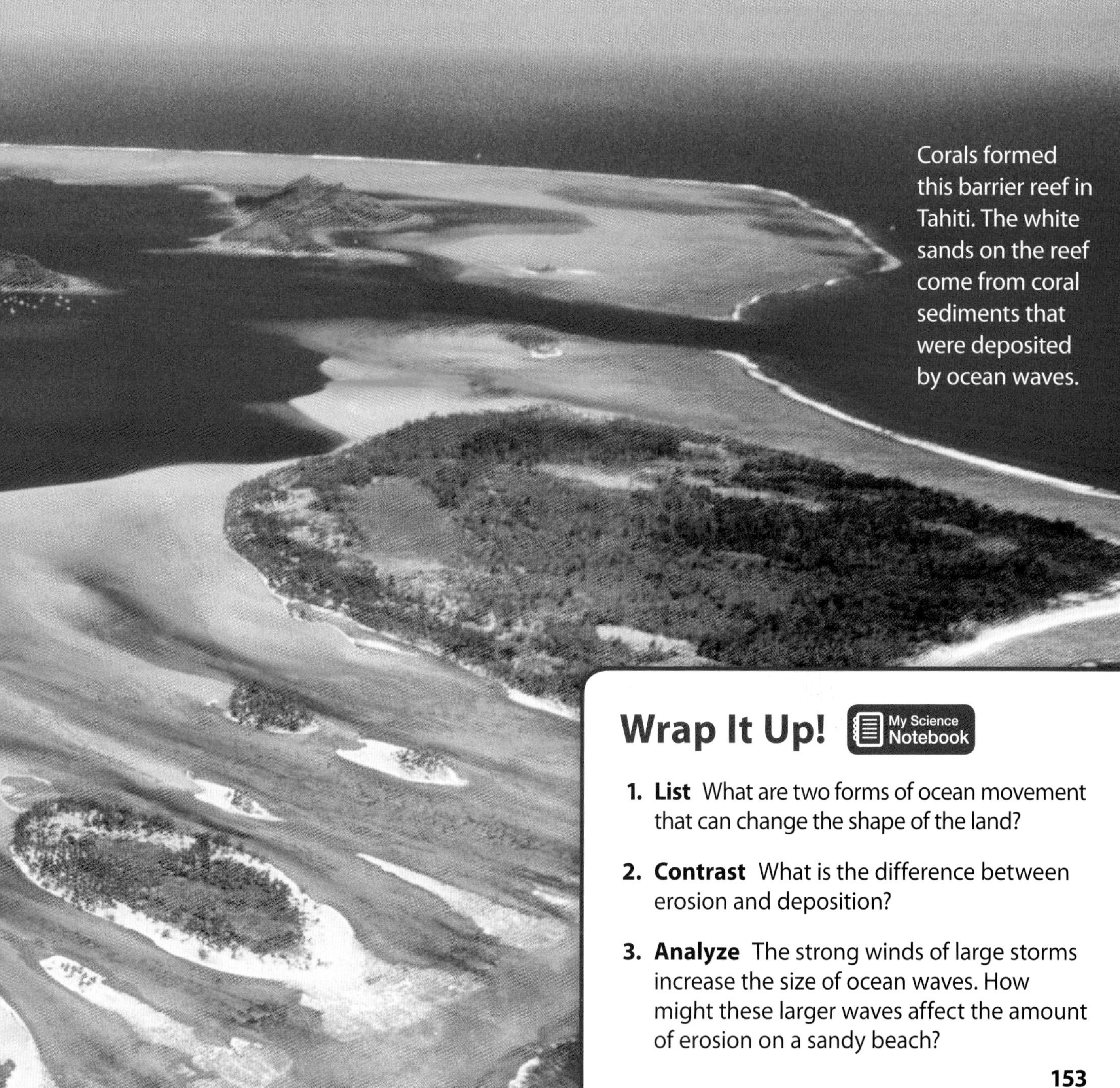

Corals formed this barrier reef in Tahiti. The white sands on the reef come from coral sediments that were deposited by ocean waves.

Wrap It Up! My Science Notebook

1. **List** What are two forms of ocean movement that can change the shape of the land?
2. **Contrast** What is the difference between erosion and deposition?
3. **Analyze** The strong winds of large storms increase the size of ocean waves. How might these larger waves affect the amount of erosion on a sandy beach?

The Ocean Affects Climate

The ocean has a strong influence on the weather and climate of coastal regions. **Weather** is the state of the atmosphere at a certain place and time. **Climate** is the pattern of the weather in an area over a long period of time.

Frequently, rain falls as moist air above the ocean moves over cooler land.

Places near the ocean usually receive more rainfall than inland areas. The ocean also reduces the extremes of high and low temperatures. This is because water warms and cools more slowly than land.

On a summer day, sunlight heats the land faster than it heats the ocean. Cool breezes blow from the ocean toward the shore. This makes the climate along the coast cooler than the climate farther inland in summer. In winter, the land cools faster than the ocean. The warmer ocean water makes the climate along the coast warmer than that of inland areas in winter.

Ocean currents also influence climate. For example, the Gulf Stream is a current that carries warm water from the Caribbean and Gulf of Mexico into the North Atlantic. Its warm water makes the climate of the East Coast of the United States and the western regions of Europe warmer than in the middle of the continents.

DCI ESS2.A: Earth Materials and Systems. Earth's major systems are the geosphere (solid and molten rock, soil, and sediments), the hydrosphere (water and ice), the atmosphere (air), and the biosphere (living things, including humans). These systems interact in multiple ways to affect Earth's surface materials and processes. The ocean supports a variety of ecosystems and organisms, shapes landforms, and influences climate. Winds and clouds in the atmosphere interact with the landforms to determine patterns of weather. (5-ESS2-1)

CCC Systems and System Models. A system can be described in terms of its components and their

In this image, reds and oranges indicate warmer temperatures. Greens and blues show cooler temperatures.

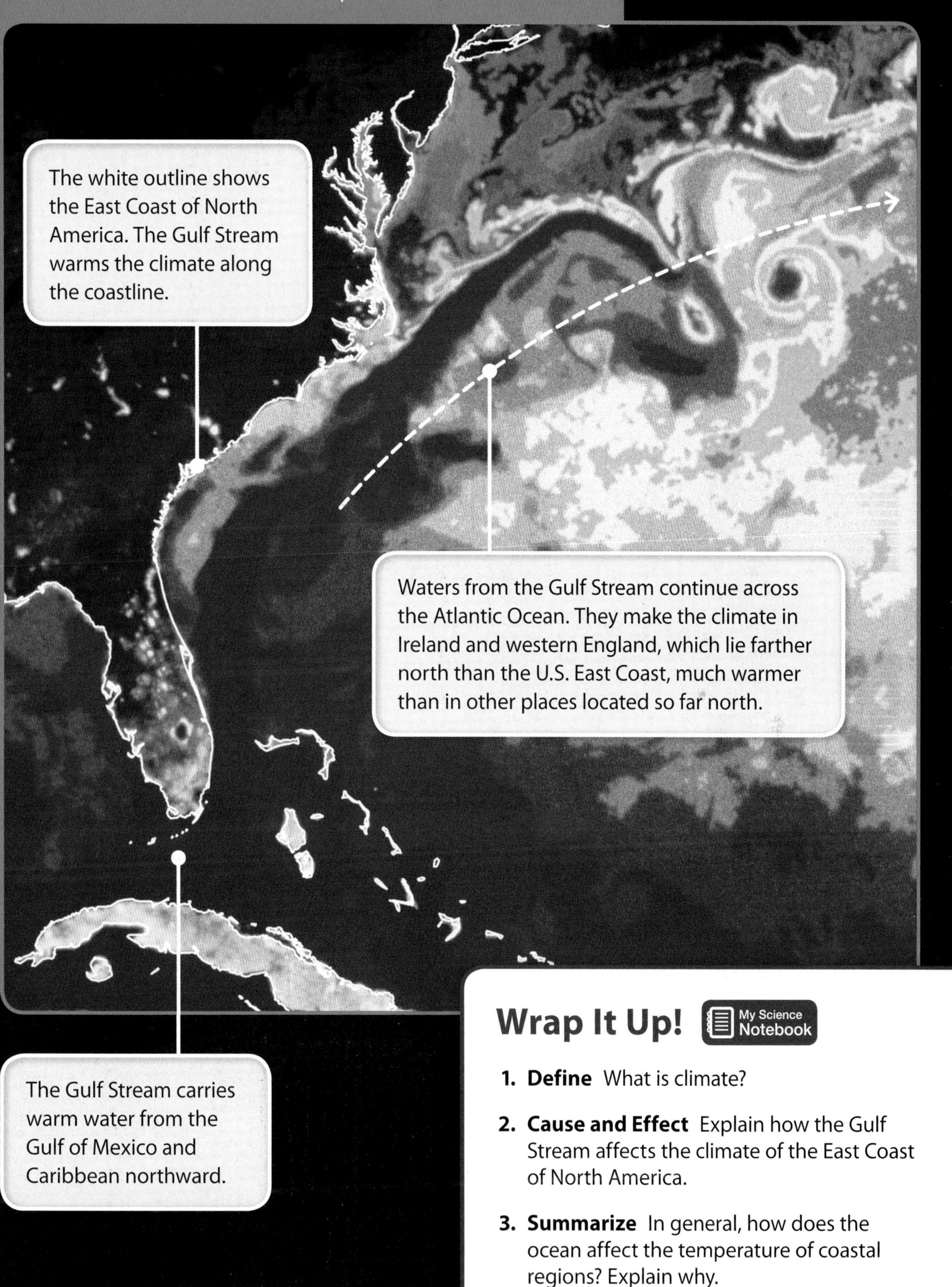

Wrap It Up! My Science Notebook

1. **Define** What is climate?
2. **Cause and Effect** Explain how the Gulf Stream affects the climate of the East Coast of North America.
3. **Summarize** In general, how does the ocean affect the temperature of coastal regions? Explain why.

Landforms and Weather Patterns

Weather changes the shape of the land. At the same time, landforms interact with winds and clouds to determine the patterns of weather in a region.

Mountain ranges affect the amount of rain that falls in an area. Winds blowing off the ocean bring moist air to the land. If there is a mountain range in the path of the wind, the air is pushed upward. As the air moves up, it cools. This cooling causes water vapor in the air to condense. Clouds form. Rain and snow from the clouds fall on the side of the mountain closest to the ocean. The air loses much of its moisture.

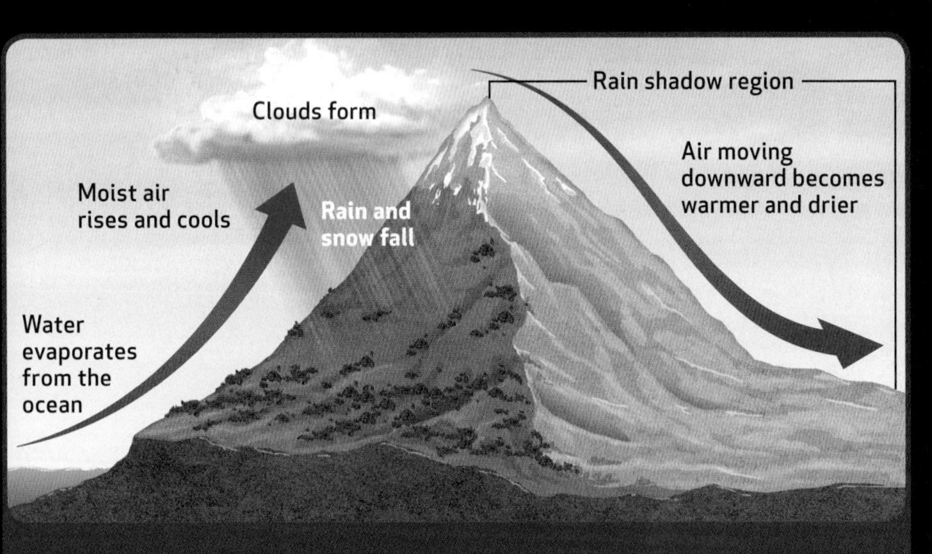

The rain shadow region forms on the side of the mountain farthest from the ocean.

DCI ESS2.A: Earth Materials and Systems. Earth's major systems are the geosphere (solid and molten rock, soil, and sediments), the hydrosphere (water and ice), the atmosphere (air), and the biosphere (living things, including humans). These systems interact in multiple ways to affect Earth's surface materials and processes. The ocean supports a variety of ecosystems and organisms, shapes landforms, and influences climate. Winds and clouds in the atmosphere interact with the landforms to determine patterns of weather. (5-ESS2-1)

CCC Systems and System Models. A system can be described in terms of its components and their interactions. (5-ESS2-1)

As the air moves down the other side of the mountain range, it gets warmer. Clouds disappear. The area on the dry side of the mountain range is said to be in a **rain shadow.** Little rain falls in these areas.

This satellite image shows a rain shadow on Mauna Kea, a volcano on the island of Hawai'i. Lush greenery grows where warm, moist air rises. Clouds form and rain falls on the coastal side of the volcano.

Wrap It Up!

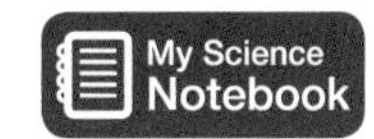

1. **Explain** Why do clouds form near the top of a mountain range?
2. **Contrast** Describe the difference between the amount of rain that falls on the ocean side of a mountain range and the amount of rain that falls in a mountain range's rain shadow.
3. **Apply** Death Valley in California and Nevada is one of the driest places in North America. Death Valley is located east of the Sierra Nevada. Why is Death Valley so dry? (*Hint*: The winds in this region blow east from the Pacific Ocean.)

The Atmosphere and Landforms

What are these strange-looking rock spires? These formations are tufa towers in Mono Lake, California. Tufa towers contain calcium carbonate, a chemical found in limestone.

Tufa towers form under water. If the water level in the lake drops, the tufa towers are exposed to the atmosphere. Then winds and rain begin to weather the towers. Strong winds that carry sediments act like sandpaper, grinding down the surface of the rocks until they eventually become smooth.

Over time, rain will dissolve the rock until these tufa towers are no longer exposed above the lake's surface.

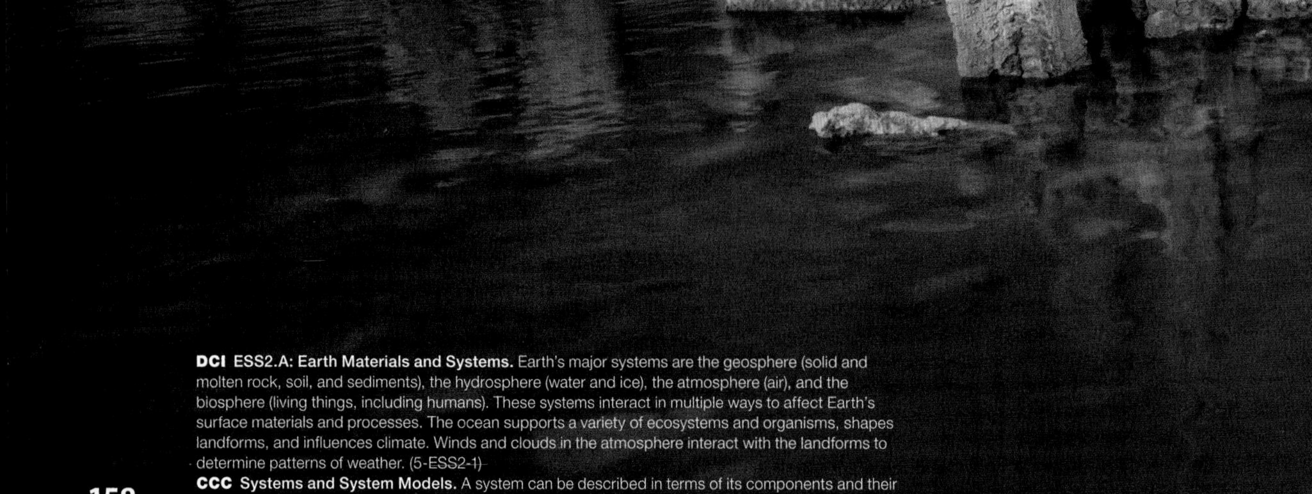

DCI ESS2.A: Earth Materials and Systems. Earth's major systems are the geosphere (solid and molten rock, soil, and sediments), the hydrosphere (water and ice), the atmosphere (air), and the biosphere (living things, including humans). These systems interact in multiple ways to affect Earth's surface materials and processes. The ocean supports a variety of ecosystems and organisms, shapes landforms, and influences climate. Winds and clouds in the atmosphere interact with the landforms to determine patterns of weather. (5-ESS2-1)

CCC Systems and System Models. A system can be described in terms of its components and their interactions. (5-ESS2-1)

Rain weathers rocks and sediments, eventually wearing down hills and even tall mountains. Rain is naturally acidic. Chemical pollution can make rain even more acidic. The acid in rain can dissolve the limestone rocks that make up the towers. The dissolved calcium carbonate runs back into the lake.

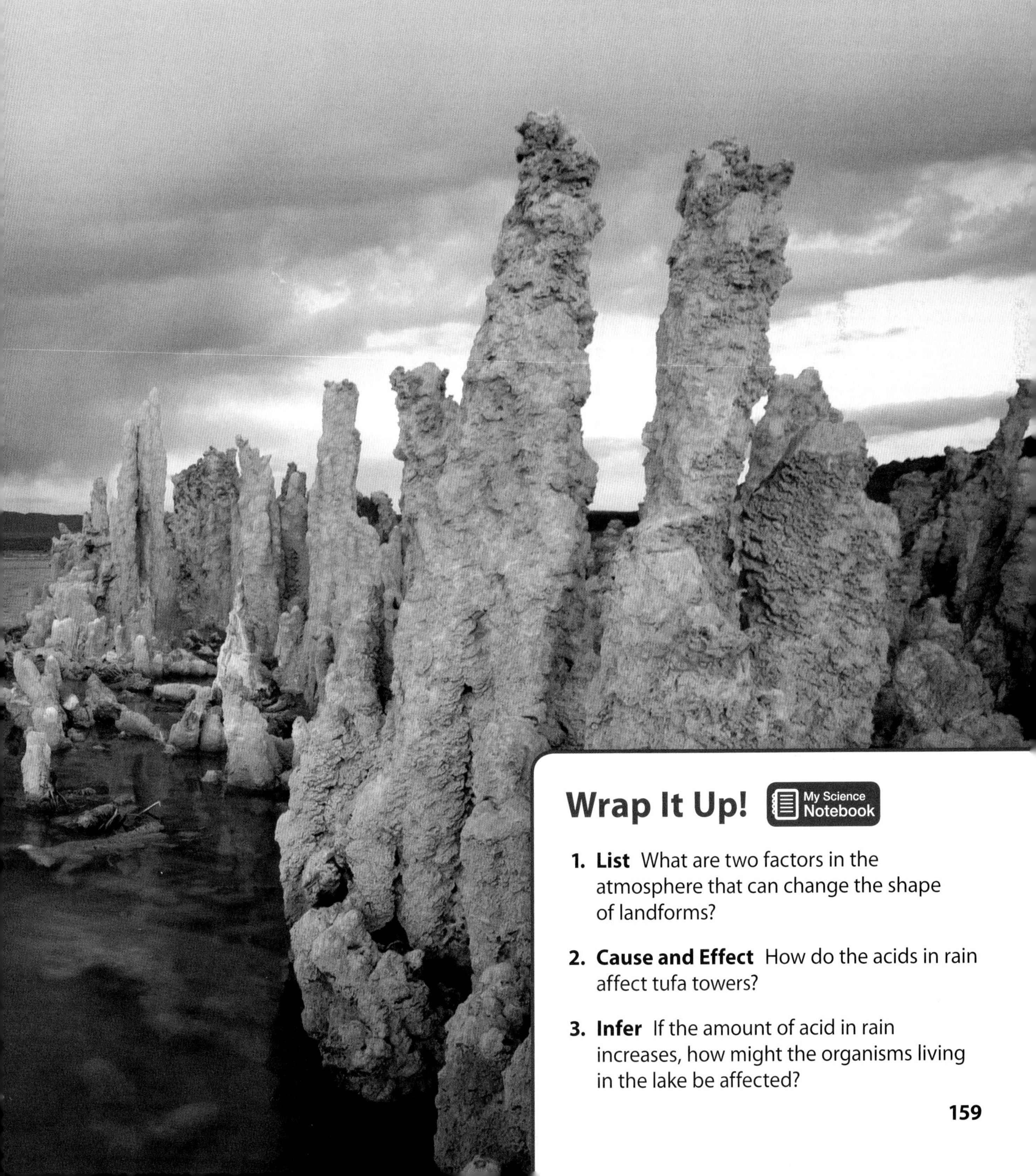

Wrap It Up! My Science Notebook

1. **List** What are two factors in the atmosphere that can change the shape of landforms?
2. **Cause and Effect** How do the acids in rain affect tufa towers?
3. **Infer** If the amount of acid in rain increases, how might the organisms living in the lake be affected?

Develop a Model

You've learned about Earth's geosphere, hydrosphere, atmosphere, and biosphere and have seen ways in which these systems interact. Now it's your turn. How can you develop a model to describe an interaction between two of Earth's spheres?

1. **Construct an explanatory model.** My Science Notebook
   Decide which interaction you will show. To get some ideas, review the lessons on Earth's spheres and ways in which they interact. You have many choices. For example, you might show how the ocean affects the shape of the land. You might show how weather and climate affect landforms, or how mountain ranges affect clouds and rain. Be sure the interaction you choose involves two *different* spheres. Draw diagrams that show the interaction.

2. **Design your model.**
   What will you use to make your model? You might make a large poster, a three-dimensional model, or even a computer animation. If your interaction causes a system to change, you could show it as a series of steps.

   Gather your materials and construct your model. Include labels that describe what is happening in each part of the interaction.

3. **Analyze and revise your model.**
   Add a third sphere and explain how it interacts with the other two spheres. Revise your models.

4. **Present your model.**
   When you are sure that your model is the best it can be, present it to the class. In your presentation, identify the spheres involved and describe their interactions.

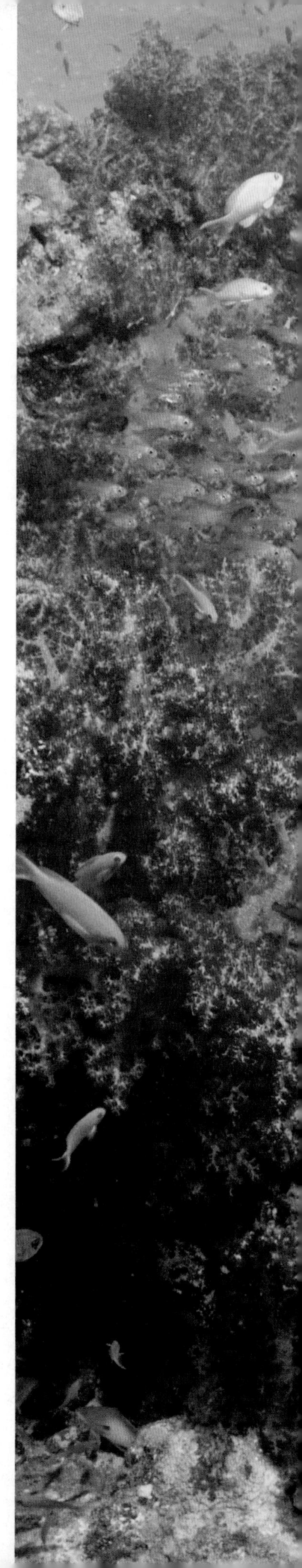

PE 5-ESS2-1. Develop a model using an example to describe ways the geosphere, biosphere, hydrosphere, and/or atmosphere interact.

All of Earth's major systems interact in this coral reef.

Water on Earth

When viewed from space, our planet looks blue. That's because most of Earth's surface is covered with water. Nearly all of Earth's water is in the ocean. Ocean water is salty. Most of Earth's fresh water is frozen in large layers of ice called **glaciers.** A smaller portion is found in groundwater, the water beneath the surface of the land. Only a tiny fraction of Earth's water is found in streams, lakes, wetlands, and the atmosphere.

Most of the fresh water on Earth is frozen in glaciers.

DCI **ESS2.C: The Roles of Water in Earth's Surface Processes.** Nearly all of Earth's available water is in the ocean. Most fresh water is in glaciers or underground; only a tiny fraction is in streams, lakes, wetlands, and the atmosphere. (5-ESS2-2)

CCC **Scale, Proportion, and Quantity.** Standard units are used to measure and describe physical quantities such as weight and volume. (5-ESS2-2)

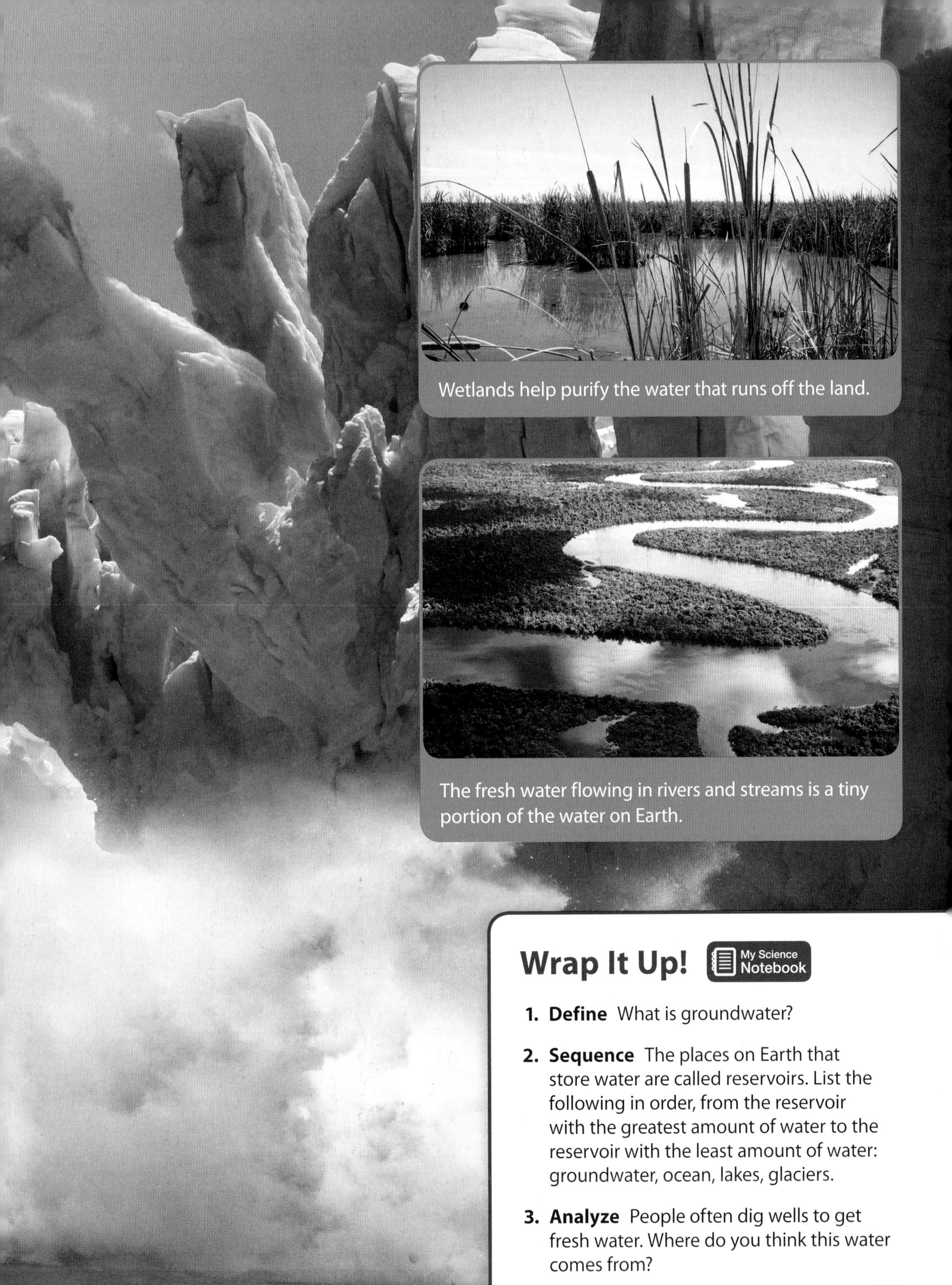

Wetlands help purify the water that runs off the land.

The fresh water flowing in rivers and streams is a tiny portion of the water on Earth.

Wrap It Up! My Science Notebook

1. **Define** What is groundwater?
2. **Sequence** The places on Earth that store water are called reservoirs. List the following in order, from the reservoir with the greatest amount of water to the reservoir with the least amount of water: groundwater, ocean, lakes, glaciers.
3. **Analyze** People often dig wells to get fresh water. Where do you think this water comes from?

Investigate

Graphing Water Data

? How does the amount of water in four of Earth's main reservoirs compare?

Most of Earth's surface is covered with water. This water is located in five main reservoirs: ocean water; ice caps and glaciers; groundwater; surface water, such as streams and lakes; and the atmosphere. In this activity, you will use data to make a graph comparing the amount of water found in four of these reservoirs.

Materials

graph paper

data chart

Reservoir	Approximate Percentage of Total Water on Earth
oceans	97%
ice caps and glaciers	2.4%
groundwater	0.4%
surface water (lakes, rivers, streams, and ponds)	$< 0.1\%$

PE 5-ESS2-2. Describe and graph the amounts and percentages of water and fresh water in various reservoirs to provide evidence about the distribution of water on Earth.

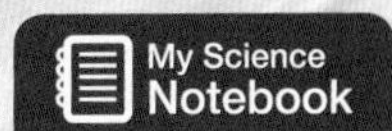

1 Examine the data. What is the largest reservoir of water? What is the smallest reservoir? Think about how you will show all the values on the chart accurately. Then use the descriptions below to help you decide which type of graph you will use to represent your data.

Sample Circle Graph

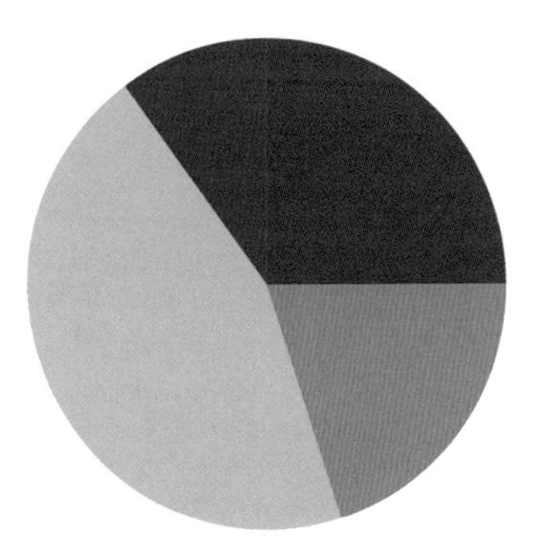

In a circle graph, the complete circle represents the total amount, or 100%. The size of each slice represents a percentage of the total.

Sample Bar Graph

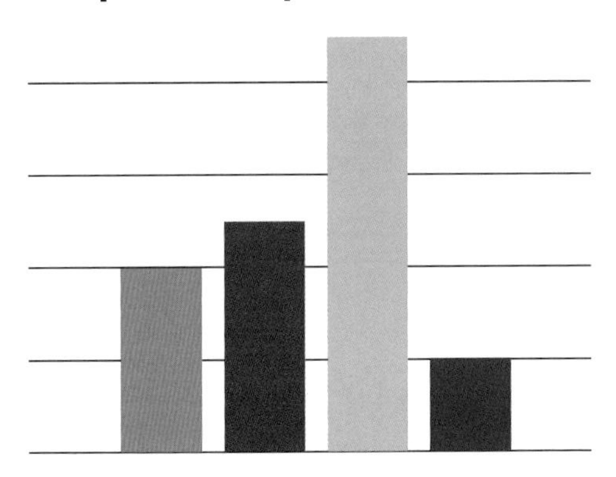

In a bar graph, each category gets a separate bar. The height of each bar indicates the amount.

2 Make your graph. Be sure to include a title and clear labels. Check to be sure that your graph represents the data accurately.

3 Share your graph with a classmate. Revise your graph based on feedback.

Wrap It Up!

1. **Describe** Use evidence from your graph to describe the distribution of salt water and fresh water on Earth.
2. **Contrast** Use your graph to describe the amount of groundwater and the amount of water that is frozen in glaciers. Which reservoir contains more water? Use evidence from your graph to support your answer.
3. **Apply** In many parts of the world there is not enough water in rivers and lakes to supply people's need for fresh water. What are some other sources of fresh water that people could use?

Earth's Resources

Earth is full of the resources that people need. People need clean air to breathe and clean water to drink. People grow plants and raise animals for food. Air, water, plants, and animals are **renewable resources,** or materials that are continually being replaced and will not run out. Use the information below to learn more about some of Earth's renewable resources.

RENEWABLE RESOURCES

Air is a renewable resource. People need clean air to breathe.

Crops such as cotton can be replanted year after year.

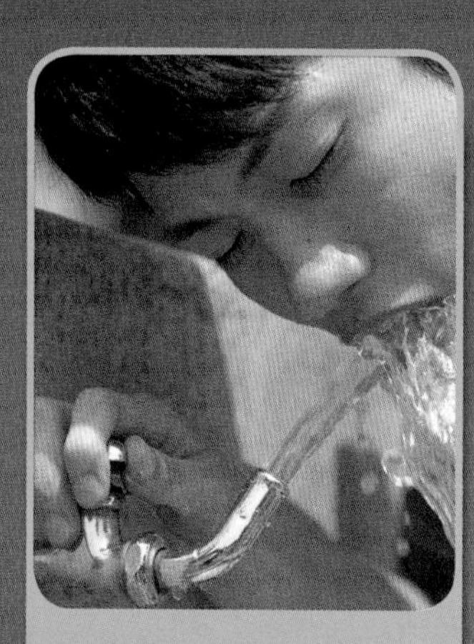

All living things need water. Water can be cleaned and used again.

Controlling the number of fish caught ensures that enough fish will live to reproduce.

DCI ESS3.C: Human Impacts on Earth Systems. Human activities in agriculture, industry, and everyday life have had major effects on the land, vegetation, streams, ocean, air, and even outer space. But individuals and communities are doing things to help protect Earth's resources and environments. (5-ESS3-1)

CCC Systems and System Models. A system can be described in terms of its components and their interactions. (5-ESS3-1)

Not all resources are renewable. **Nonrenewable resources** are those that cannot be replaced quickly enough to keep from running out. Materials that form very slowly are nonrenewable. Coal, oil, and natural gas formed slowly from the remains of plants and animals that lived long ago. These resources are nonrenewable. Use the information below to learn more about some of Earth's nonrenewable resources.

NONRENEWABLE RESOURCES

Coal formed slowly over time from plants that lived millions of years ago. This resource could run out.

People use oil at a fast pace. The oil they use cannot quickly be replaced.

Natural gas is burned to heat water and homes. The gas that people use cannot quickly be replaced.

Wrap It Up!

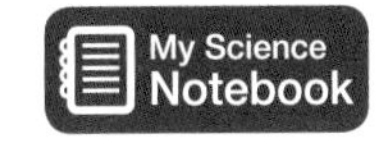

1. **Contrast** What is the difference between renewable and nonrenewable resources?
2. **Classify** Are trees a renewable or nonrenewable resource? Explain your answer.
3. **Apply** List three ways you use renewable resources and three ways you use nonrenewable resources.

Humans Impact the Land

Human activities affect the land in many ways. These activities include **agriculture,** or farming; the many activities of everyday life that require buildings and roads; and industry.

When farmers plow the land, they expose the soil to wind and rain. This can cause erosion, or the washing away of soil. In dry regions people often dig ditches or use spray systems to irrigate the land.

By planting crops along the contour of the land, farmers can reduce soil erosion.

DCI ESS3.C: Human Impacts on Earth Systems. Human activities in agriculture, industry, and everyday life have had major effects on the land, vegetation, streams, ocean, air, and even outer space. But individuals and communities are doing things to help protect Earth's resources and environments. (5-ESS3-1)

CCC Systems and System Models. A system can be described in terms of its components and their interactions. (5-ESS3-1)

People use large amounts of land to build homes and cities. They also build roads, railroad tracks, and airports so they can travel from place to place. These activities change where plants can grow and where water runs.

People also take many materials from the ground to use in industry. Some miners dig huge pits to remove materials such as coal, copper, and gravel. Mining activities may also pollute the land and water.

People can plan to reserve spaces without buildings and roads for vegetation and wildlife.

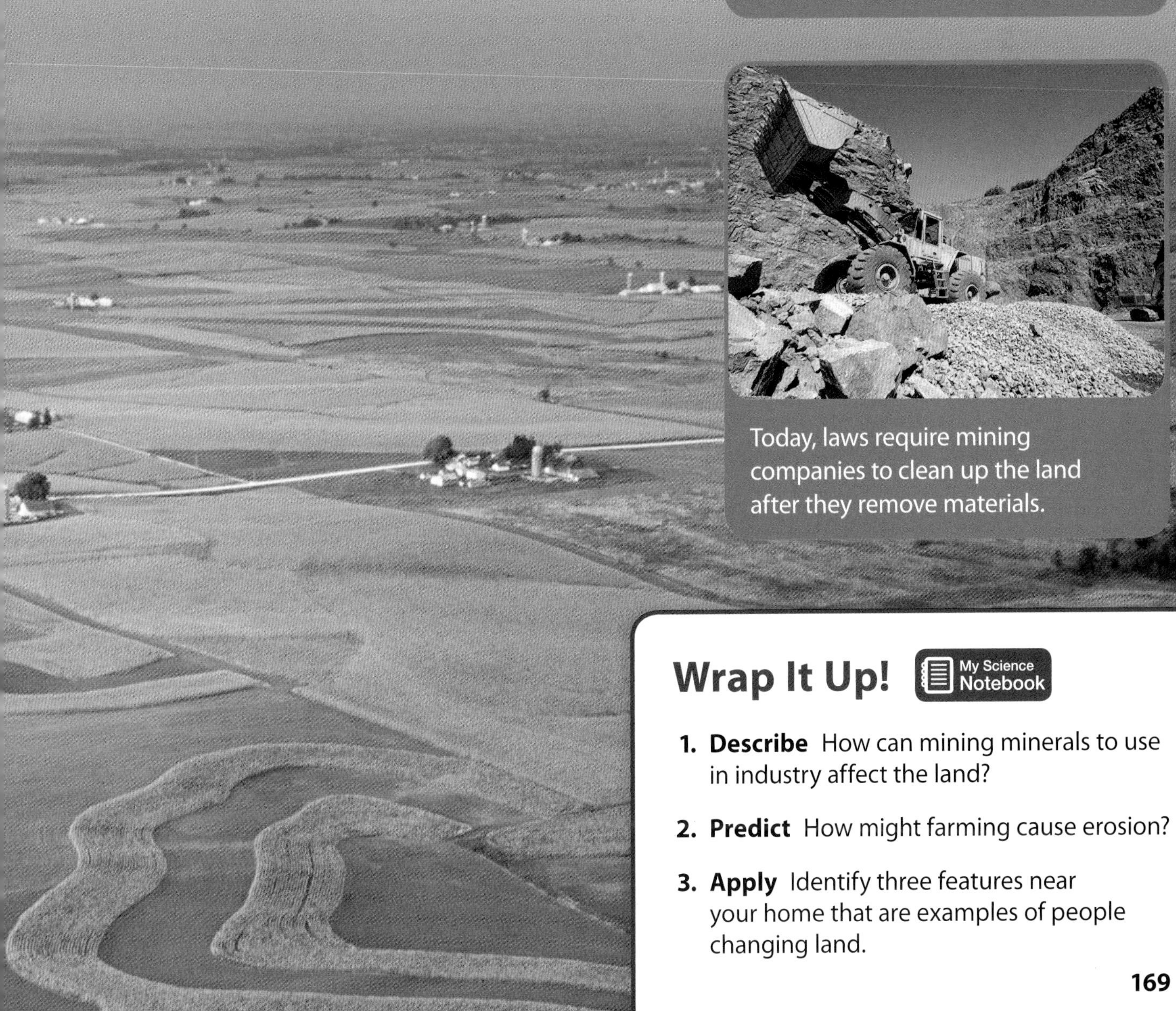

Today, laws require mining companies to clean up the land after they remove materials.

Wrap It Up! My Science Notebook

1. **Describe** How can mining minerals to use in industry affect the land?
2. **Predict** How might farming cause erosion?
3. **Apply** Identify three features near your home that are examples of people changing land.

Humans Impact Vegetation

Human activities can have a major effect on the **vegetation,** or the plants growing in an area. Farmers often cut down trees or burn forests so they can grow crops or raise livestock. Other people cut down trees to sell the wood or to make paper. The removal of all of the trees in an area is called **deforestation.** Without trees, there are no roots to hold the soil in place. Water and wind carry the soil away.

The soil in many prairies and other grasslands is very fertile. Because the soil is good for agriculture, most of the prairies in the United States have been converted into cropland. Corn, wheat, and soybeans now grow where tall grasses and wildflowers once grew.

Cutting down trees can leave soil exposed to erosion.

DCI ESS3.C: Human Impacts on Earth Systems. Human activities in agriculture, industry, and everyday life have had major effects on the land, vegetation, streams, ocean, air, and even outer space. But individuals and communities are doing things to help protect Earth's resources and environments. (5-ESS3-1)

CCC Systems and System Models. A system can be described in terms of its components and their interactions. (5-ESS3-1)

When humans build new houses, they usually remove the vegetation. Humans also remove vegetation so they can build cities, factories, and roads.

Certain human activities produce pollution that can damage or even kill plants. For example, some factories and electric power stations release acidic chemicals that contaminate the rain and clouds. When this acid rain falls to the ground, it can kill trees and other plants.

Some chemicals make rain more acidic. Acid rain can kill trees and other vegetation.

Wrap It Up! My Science Notebook

1. **Define** What is deforestation?
2. **Explain** How does agriculture affect the vegetation of grasslands?
3. **Make Judgments** Some people live in suburbs where houses have large yards and gardens. Other people live in tall buildings in cities. Which kind of home do you think has a greater impact on the land? Explain.

Investigate

Plants and Pollution

? How does vinegar affect the growth of plants?

Some human activities pollute the land and water. Many factories and power plants release acidic chemicals, which can enter the air and water. Other activities add salt to the land. For example, people sometimes put salt on roads to melt ice. In this activity, you'll investigate how acidic chemicals affect the growth of plants.

Materials

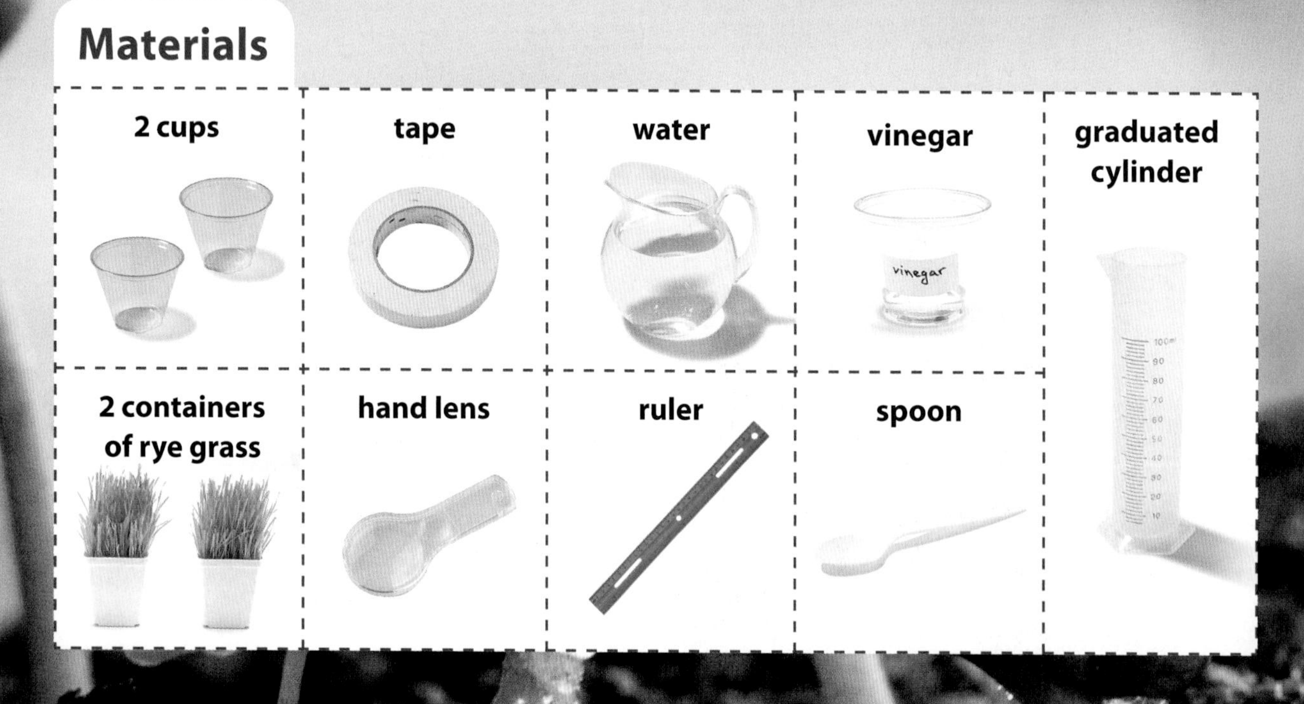

DCI ESS3.C: Human Impacts on Earth Systems. Human activities in agriculture, industry, and everyday life have had major effects on the land, vegetation, streams, ocean, air, and even outer space. But individuals and communities are doing things to help protect Earth's resources and environments. (5-ESS3-1)

CCC Systems and System Models. A system can be described in terms of its components and their interactions. (5-ESS3-1)

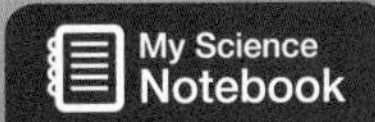

1 Label the cups as shown. Use the graduated cylinder to pour 75 mL of water into the *clean water* cup. Then use the graduated cylinder to add 25 mL of vinegar and 50 mL of water to the *vinegar and water* cup.

2 Label one container of rye grass *clean water.* Label the other container *polluted water.* Use a hand lens to observe the grass in each container. Use the ruler to measure the height of the grass in each container. Record your measurements and other observations.

3 Add two spoonfuls of clean water to the grass in the *clean water* container. Add two spoonfuls of water and vinegar to the grass in the *polluted water* container.

4 Repeat step 3 every day for two weeks. Observe and measure the grass in each container every other day. Record your observations and measurements.

Explore on Your Own

What would happen if you repeat the investigation using a different form of pollution, such as salt water? Plan and carry out your own investigation. Record your observations. Compare the results of your investigations.

Wrap It Up!

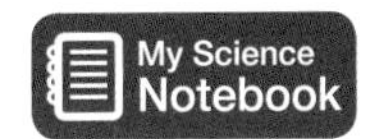

1. **Compare and Contrast** How is the growth of the rye grass in the cups alike and different? How does vinegar affect the growth of rye grass? Use data from your observation in your answer.
2. **Apply** How could you plan an investigation to see how salt affects the growth of rye grass?
3. **Research** What else would you like to find out about how pollution in water affects the growth of plants? How could you find out?

Humans Impact Water

Over the centuries, human activities have been closely connected with the supply of fresh water. Since ancient times, farmers have removed water from rivers and streams to irrigate their crops. More recently, people have dammed streams and rivers to increase the supply of drinking water, to power mills, and to generate electricity. Today human activities use more than half of the fresh water flowing in Earth's streams and rivers.

Wastes from homes, farms, mines, and factories can end up in water. Fertilizers put on fields and lawns can run off the land and into streams and lakes. Water and pollutants that flow off the land are called **runoff.** Detergents and wastes from households can also flow into streams and

DCI **ESS3.C: Human Impacts on Earth Systems.** Human activities in agriculture, industry, and everyday life have had major effects on the land, vegetation, streams, ocean, air, and even outer space. But individuals and communities are doing things to help protect Earth's resources and environments. (5-ESS3-1)

CCC **Systems and System Models.** A system can be described in terms of its components and their interactions. (5-ESS3-1)

lakes. These pollutants cause the number of algae growing in streams, rivers, and lakes to increase. The algae use up the oxygen in the water, killing fish and other organisms.

Pollutants and trash that enter streams and rivers eventually reach the ocean. Where some rivers flow into the ocean, pollutants produce a dead zone where few organisms can live. Oil and other materials that are carried by ships can also pollute the ocean.

Human trash often ends up in the ocean. There it can harm fish, birds, and other organisms.

When people build dams to produce electricity, a lake forms where there was once a flowing stream or river. This can affect the plants and animals that live in the water.

Wrap It Up!

1. **Define** What is runoff?
2. **Explain** What are two ways that agriculture can affect Earth's freshwater supply?
3. **Apply** List three ways that you and your family use water every day. How could you reduce pollution in your everyday water use?

Humans Impact Air

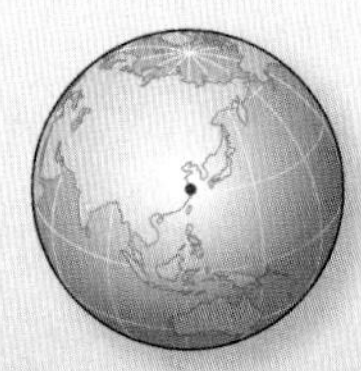

Many human activities affect the quality of the air. For example, when people burn fuels such as wood and charcoal to heat their homes, tiny particles of ash are released into the air. Factories and power plants that burn coal and oil release many chemicals, including carbon dioxide. Some of these chemicals combine with water in the air to form acid rain.

In many cities, including Shanghai, China, shown here, the main source of air pollution is cars, trucks, and buses that burn gasoline and diesel fuel. These vehicles release chemicals that react with sunlight to form a brown haze called **smog.**

China has some of the poorest air quality in the world. The smog is visible in this photo of Shanghai.

DCI ESS3.C: Human Impacts on Earth Systems. Human activities in agriculture, industry, and everyday life have had major effects on the land, vegetation, streams, ocean, air, and even outer space. But individuals and communities are doing things to help protect Earth's resources and environments. (5-ESS3-1)

CCC Systems and System Models. A system can be described in terms of its components and their interactions. (5-ESS3-1)

Today many people are working together to clean up the air. Most factories in the United States now have devices in their smokestacks that remove some of the pollution before it is released into the air. Planting trees and rooftop gardens in cities also helps reduce air pollution.

This rooftop garden not only helps clean the air, it also provides fresh food for a restaurant.

Wrap It Up!

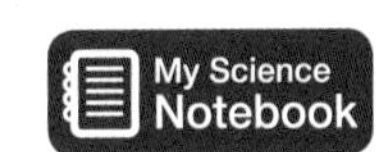

1. **Recall** What are three ways that burning fuels affects the quality of the air?
2. **Explain** What are some ways that people are working together to reduce air pollution?
3. **Apply** Instead of riding in a car, you decide to ride your bicycle to school. How could this decision affect air quality? Explain.

Humans Impact Space

You might think that the space around Earth is empty. But instead, it contains millions of pieces of human-made trash—**space junk**! For more than 50 years, humans have been sending satellites into orbit around Earth. When satellites stop working, they become space junk. In addition to old satellites, space junk includes pieces of rockets, metal shards from objects that collided in space, and even tools left behind by astronauts.

Space junk is still in orbit but is no longer controlled by people on Earth. The objects are moving rapidly—27,400 kilometers per hour (17,000 miles per hour) or more! These objects can cause serious damage if they hit a satellite, spacecraft, or an astronaut.

Eventually Earth's gravity pulls bits of space junk back into Earth's atmosphere. Most of them burn up before reaching Earth's surface. Space junk that reaches the surface usually falls into the ocean. No one has ever been injured by space junk falling to Earth.

Today space scientists are working to reduce the amount of trash that is left in space. They are using radar to monitor the location of large objects. They use this information to help spacecraft, such as the International Space Station, avoid collisions with space junk.

DCI ESS3.C: Human Impacts on Earth Systems. Human activities in agriculture, industry, and everyday life have had major effects on the land, vegetation, streams, ocean, air, and even outer space. But individuals and communities are doing things to help protect Earth's resources and environments. (5-ESS3-1)

CCC Systems and System Models. A system can be described in terms of its components and their interactions. (5-ESS3-1)

Most space junk that falls to Earth is about the size of a walnut, such as the pieces on the left. Some of the space junk, however, is huge. This large piece is thought to be a fuel tank from a Delta 2 rocket.

This computer-generated image shows the Earth-orbiting objects that NASA is tracking. Most of these objects are space junk, not working satellites. The dots do not represent the size of the objects. They are enlarged to make them easier to track.

Wrap It Up!

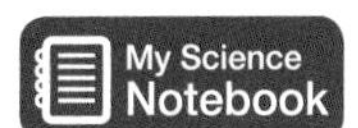

1. **Describe** What is space junk? Give some examples.
2. **Cause and Effect** Why might space junk be dangerous?
3. **Identify** How are scientists working to help spacecraft avoid collisions with space junk?

Protecting Land, Air, and Water

Individuals and communities are doing many things to help protect Earth's resources and environment. The careful use and protection of natural resources is known as **conservation.** Conservation is important. If nonrenewable resources such as oil and coal get used up, they will be gone forever.

HOW YOU CAN CONSERVE

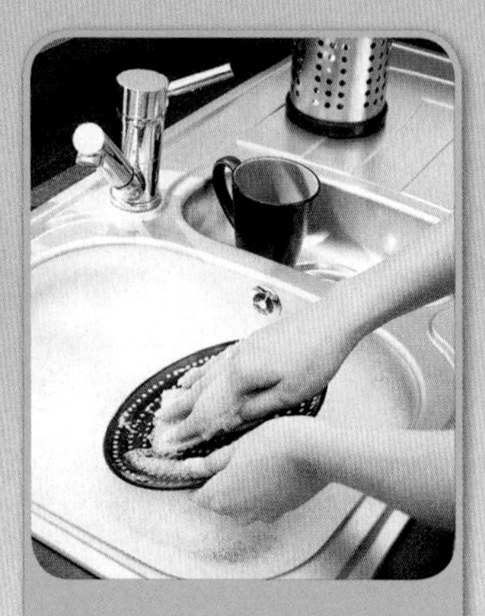

Turn off the water. Wash all the dishes, and *then* rinse to save water.

Compost. Using compost from table scraps returns nutrients to the soil.

Use alternative transportation. Riding a bike is good for the air and for your body.

Recycle. Recycling reduces the need for new resources to be mined or refined.

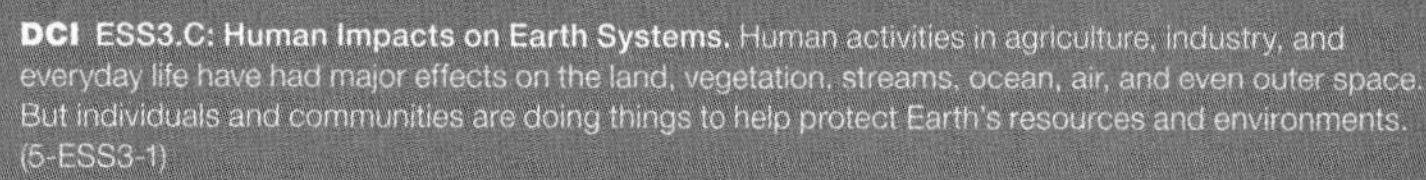

DCI ESS3.C: Human Impacts on Earth Systems. Human activities in agriculture, industry, and everyday life have had major effects on the land, vegetation, streams, ocean, air, and even outer space. But individuals and communities are doing things to help protect Earth's resources and environments. (5-ESS3-1)

CCC Systems and System Models. A system can be described in terms of its components and their interactions. (5-ESS3-1)

Even renewable resources need to be used carefully. For example, when companies cut down forests for lumber, they often plant new trees to replace the ones that they cut down. People also **recycle,** or use materials again.

Today many communities are setting aside land to conserve ecosystems. Cities are reducing air pollution by building public transportation systems so people do not have to drive cars to work.

Aluminum is one of the easiest materials to recycle.

SCIENCE in a SNAP

Recycle by the Numbers

1 Recycling is one way to conserve natural resources. Numbers on plastic products show the kind of plastic from which the objects are made. Most plastics that can be recycled have a 1 or 2 on them.

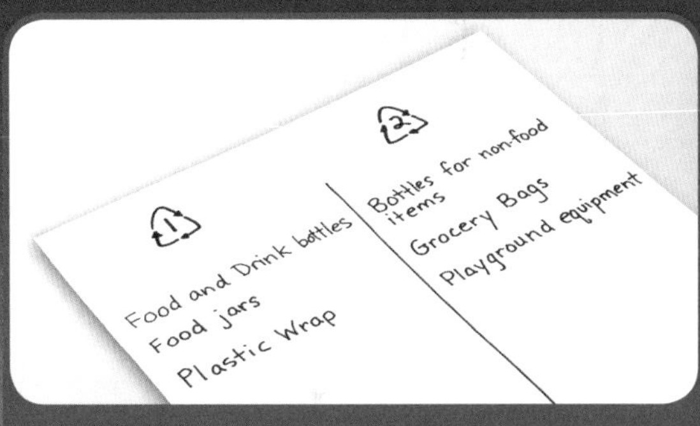

2 Look for numbers on plastic objects. List them as 1, 2, or other numbers.

? **What number was on most of the objects? What other numbers did you find?**

Wrap It Up!

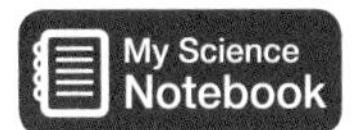

1. **Define** What is recycling?
2. **Apply** Give three examples of things you can do to conserve natural resources.
3. **Evaluate** Trees are a renewable resource. Why is it important to conserve forests? Explain your answer.

Environmental Pioneer

DCI **ESS3.C: Human Impacts on Earth Systems.** Human activities in agriculture, industry, and everyday life have had major effects on the land, vegetation, streams, ocean, air, and even outer space. But individuals and communities are doing things to help protect Earth's resources and environments. (5-ESS3-1)
CCC **Systems and System Models.** A system can be described in terms of its components and their interactions. (5-ESS3-1)
NS **Science Addresses Questions About the Natural and Material World.** Science findings are limited to questions that can be answered with empirical evidence. (5-ESS3-1)

Rachel Carson intended *Silent Spring* to be a warning to people about the misuse of pesticides.

Rachel Carson is most famous for her book, *Silent Spring,* in which she challenged the use and effects of pesticides in agriculture. This created controversy that continues today. Rachel grew up in Pennsylvania and developed a love of nature early in life. She studied biology and received a master's degree in zoology from Johns Hopkins University in 1932.

She was always writing about nature and science. She won her first award for writing at the age of 11. After graduating from university, she worked for the U.S. Bureau of Fisheries writing radio scripts about marine life. She became the editor-in-chief of publications for the U.S. Fish and Wildlife Service.

While Rachel was working for the federal government, she also wrote her own books. One of the themes of her books was that people were part of any given ecosystem. They could impact it with their actions. Her books were so popular that she left her job in 1952 to work on her writing full time. Rachel's book *Silent Spring* was published in 1962, and focused strongly on the theme of human impact on ecosystems.

Rachel researched the connection between the increased use of insecticides after World War II and the decrease in organisms such as fish and birds. These animals depend on insects as a source of food. In *Silent Spring,* Rachel was particularly critical of the use of DDT to control mosquitoes and other insect pests. DDT is a human-made pesticide that was used during World War II to help reduce the spread of malaria, which is spread by mosquitoes. After the war, DDT was widely used to control many insect pests.

In *Silent Spring,* Rachel detailed connections among Earth's systems. She explained that airborne pesticides made their way into water, soil, and living things. She explained how the toxins in DDT could build up in the tissues of organisms. It is then passed along through a food chain. One example from her book included American robins dying of insecticidal poisoning. The robins ate earthworms that had high levels of DDT in their tissues. The earthworms had accumulated the DDT from eating leaves that had fallen from trees sprayed with the insecticide. Rachel's book contained other examples of the harmful impacts of DDT on humans. These were all supported by scientific studies.

Rachel Carson intended *Silent Spring* to be a warning to people about the misuse of pesticides. The book resulted in a firestorm of controversy from all sides—farmers, biologists, government officials, pesticide manufacturers, and the public. As a result of her work on *Silent Spring* and the heated debates it caused, Rachel is considered by some to be the founder of the modern-day environmentalist movement. Since her death in 1964, Rachel's advocacy for protecting the environment, including the humans that live in it, has continued to influence and inspire generations of readers.

Wrap It Up!

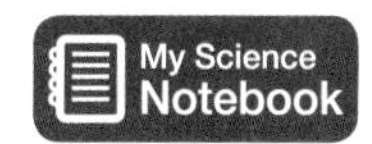

1. **Explain** How did DDT enter the bodies of organisms, such as fish and birds?
2. **Apply** Why do you think *Silent Spring* and other books like it are important?

Tower of Trees

Problem

How can people grow more trees in cities when there is no more space available?

Imagine a landscape of brick, concrete, steel, and asphalt. That's what the landscape in many cities looks like. Why? Because there are few, if any, trees! Trees provide many benefits to a city. Their leaves absorb pollutants through openings in the leaves' surfaces. Particulates, or tiny particles of pollution, are trapped and filtered by leaves, stems, and twigs. These particulates, which include dust, ash, and smoke, can cause illness. Tree leaves absorb carbon dioxide and give off oxygen. Trees help cool the air as water evaporates from their leaves. They provide cooling shade. And they look beautiful, too!

Few people would argue that trees aren't good for cities. The problem is finding a place to put them. Unless a city was planned with parks and green space in mind, it can be hard to find places for large plants such as trees.

DCI ESS3.C: Human Impacts on Earth Systems. Human activities in agriculture, industry, and everyday life have had major effects on the land, vegetation, streams, ocean, air, and even outer space. But individuals and communities are doing things to help protect Earth's resources and environments. (5-ESS3-1)

CCC Systems and System Models. A system can be described in terms of its components and their interactions. (5-ESS3-1)

There is little space in crowded cities for mature trees or even the grassy lawns of the suburbs.

Solution

Planting Trees on Buildings

One idea for increasing the number of trees in cities is to build **vertical forests**—tall buildings on which trees and other plants are grown. The first vertical forest is under construction in the city of Milan, Italy.

Engineers and scientists hope that a vertical forest will provide many of the benefits of a natural forest for the people living in it. These benefits could include cleaner air, cooler summer temperatures, less noise, and a beautiful view. The trees will be watered using recycled water from the building's baths, sinks, washing machines, and dishwashers.

Some people worry that the towers will be too hot and windy for trees to grow. But nearly everyone is hoping that the vertical forest will be a success!

Planting trees on the outside of a tall building takes some careful engineering!

This architect's drawing shows over 700 trees and thousands of shrubs and small flowering plants that will be in place on the first vertical forest being built in Milan, Italy.

Wrap It Up!

My Science Notebook

1. **Describe** What is a vertical forest?
2. **Explain** How could a vertical forest improve air quality?
3. **Evaluate** Do you think vertical forest towers are a good solution for your community? Explain why or why not.

Renewable Energy Resources

What are the sources of the energy you use? Most people in the United States rely on gasoline to run their cars. Many use oil or natural gas to heat their homes. Much of the electricity you use may come from power plants that burn coal. But gasoline, oil, coal, and natural gas are **nonrenewable energy resources.** They are being used far more quickly than they are being produced. Burning these fuels also produces air pollution.

Solar Energy Solar energy, or energy from the sun, is a clean, renewable energy resource. Solar cells use energy from sunlight to produce electricity. Some solar panels contain thousands of solar cells.

DCI ESS3.C: Human Impacts on Earth Systems. Human activities in agriculture, industry, and everyday life have had major effects on the land, vegetation, streams, ocean, air, and even outer space. But individuals and communities are doing things to help protect Earth's resources and environments. (5-ESS3-1)

CCC Systems and System Models. A system can be described in terms of its components and their interactions. (5-ESS3-1)

More and more of the energy people currently use comes from **renewable energy resources,** which are continually replaced and will not run out. These resources include solar energy, wind energy, and hydroelectric power. The use of renewable energy resources helps conserve nonrenewable resources. It also produces far less pollution than burning oil, coal, or natural gas.

Wind Energy The energy in wind can be used to produce electricity. The wind spins the wing-like blades of the turbines. Their motion runs generators, the machines that produce electricity. The electricity travels through wires to homes and businesses.

Hydroelectric Energy The energy of moving water can be used to generate electricity. When water from a lake flows through the dam, it turns large turbines. The spinning turbines run generators, which produce electricity.

Wrap It Up!

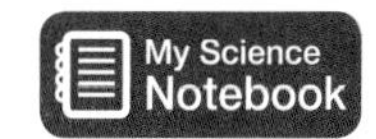

1. **Identify** What are three renewable sources of energy?
2. **Cause and Effect** How does a hydroelectric power plant produce electricity?
3. **Summarize** How does the use of renewable energy help protect all of Earth's resources?

STEM
SCIENCE
TECHNOLOGY
ENGINEERING
MATH

RESEARCH PROJECT

Energize!

Have you ever heard of wind and solar farms? No, they are not where wind and sunlight are grown. But in a way, it is where wind and solar energy are "harvested." A wind farm is a group of wind turbines. The turbines spin in the wind and generate electricity. Solar farms are areas where many solar panels convert sunlight to electricity. Wind and solar farms are becoming more common. In many areas, costs for wind and solar energy have come down. These energy sources compete with energy that is generated from fossil fuels, such as coal, oil, and natural gas.

Some people are going even further than relying on wind and solar farms. They are beginning to use wind and solar power at their homes. New product designs are making it easier to use these renewable energy resources. All energy sources have advantages and disadvantages. However, unlike energy from fossil fuels, solar and wind energy do not contribute to air and water pollution.

DCI ESS3.C: Human Impacts on Earth Systems. Human activities in agriculture, industry, and everyday life have had major effects on the land, vegetation, streams, ocean, air, and even outer space. But individuals and communities are doing things to help protect Earth's resources and environments. (5-ESS3-1)

SEP Obtaining, Evaluating, and Communicating Information. Obtain and combine information from books and/or other reliable media to explain phenomena or solutions to a design problem. (5-ESS3-1)

CCC Systems and System Models. A system can be described in terms of its components and their interactions. (5-ESS3-1)

This solar farm focuses sunlight on a boiler tower. Super hot water vapor spins turbines to produce electricity.

The Challenge

Your challenge is to work with a partner to identify one product each for both wind and solar energy that you would like to learn more about. You will need to think of one or more questions for each topic. Do research to find answers to your questions. Then you will make a booklet, poster, or computer slide demonstration to present to your class.

STEM
RESEARCH PROJECT
(continued)

1 Select a topic.

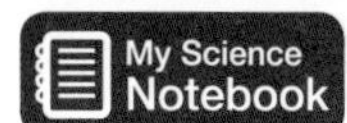

Work with your partner to brainstorm a list of questions you might want to research. Think about how you have seen wind or solar energy being used. It might be on a large scale, such as a wind or solar farm. It might be smaller-scale generators designed for home use. You might research cutting-edge systems and products. You might research new ways people are making use of wind and solar energy products.

Write down any ideas that come to you. Research and report on one wind energy product and one solar energy product. Narrow down your choices. What are some questions you have about these? Record your questions. Identify key words in your questions. Use the key words to help guide your research.

2 Plan and conduct research.

Make a plan with your partner to do research. Your teacher will help you find online and printed resources that have accurate information. In addition to the questions you wrote in Step 1, research the answers to the following questions for each product:

- How does the product work?
- How and where is the product manufactured?
- What are the costs involved in manufacturing?
- What are the advantages and disadvantages of the product?
- What are two interesting facts about the product?

Record the information you find, including the source. Use outlining or graphic organizers to keep your notes organized.

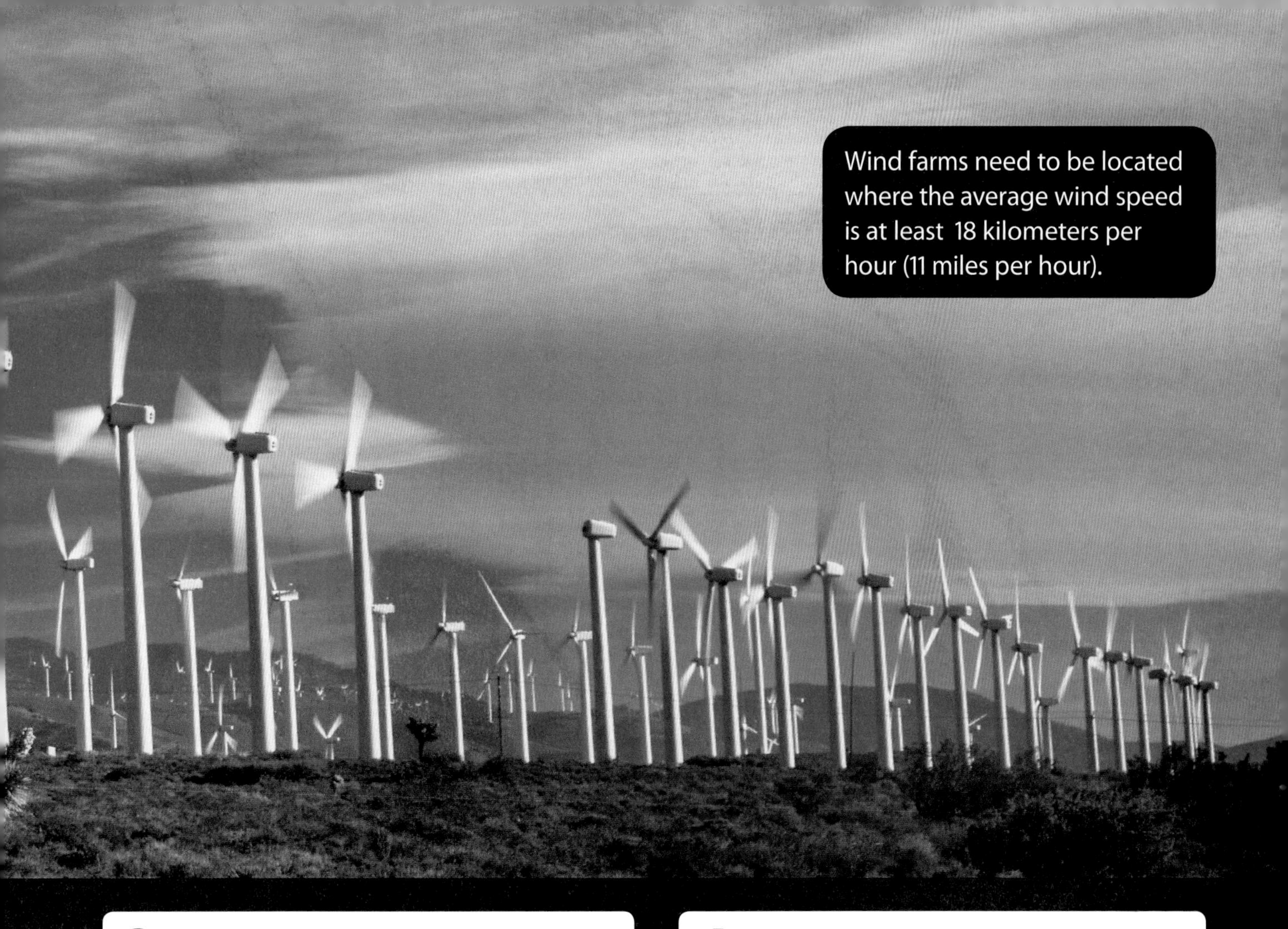

3 Draft your report.

Your report will be in the form of a booklet, poster, or computer slide presentation. In your report, summarize the main ideas. The information you present should be in your own words. Your report should include the following for each product:

- the name and picture of the product
- an explanation of how the product works or what it does
- a description of who can use the product and the advantages and disadvantages of it
- a description of how the product affects the environment
- two interesting facts about the product

Review and revise the draft of your report to make it the best it can be. Do more research to add additional information as needed. Make the final draft of your booklet, poster, or computer slide presentation.

4 Present your report.

Work with your partner to decide who will give each part of the presentation. Decide how you will display visual information.

Practice giving your part of the report aloud. Your oral presentation should express main ideas that are supported by significant details. Ask your partner to give you feedback you can use to make your part of the presentation better.

With your partner, present your report to the class. Put information in a logical order. Use descriptions, facts, and details to describe the wind and solar products. Remember to give two additional interesting facts about each product.

Listen as your classmates present their reports. How many different wind and solar energy products did your classmates identify and report on?

Investigate

Using Solar Energy

? How can you use energy from the sun to make water cleaner?

In many parts of the world there is little fresh water for people to use. In some places, people use solar energy to produce clean, fresh water from salt water or polluted water. In this activity you will investigate one way this can be done.

Materials

plastic container	clay	plastic cup	measuring cup	water
sandy soil  	spoon	plastic wrap  	rubber band 	rock 

DCI ESS3.C: Human Impacts on Earth Systems. Human activities in agriculture, industry, and everyday life have had major effects on the land, vegetation, streams, ocean, air, and even outer space. But individuals and communities are doing things to help protect Earth's resources and environments. (5-ESS3-1)

CCC Systems and System Models. A system can be described in terms of its components and their interactions. (5-ESS3-1)

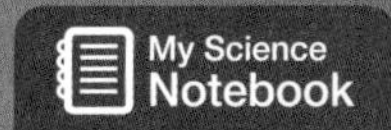

1 Use the clay to stick the plastic cup to the center of the container. Add 500 mL of water to the container. Do not put any water into the plastic cup.

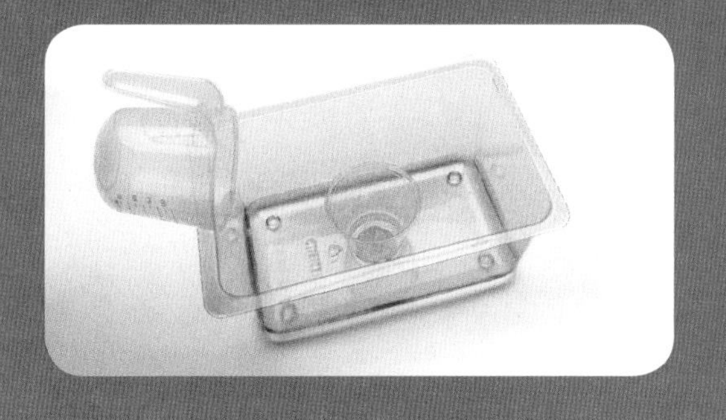

2 Add several spoonfuls of soil to the water in the container. Use the spoon to gently stir the water. Observe the plastic cup and the container. Record your observations.

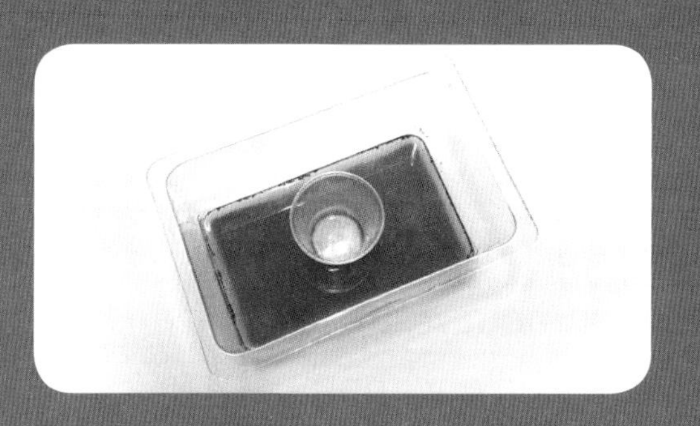

3 Cover the container with plastic wrap. Use a rubber band to seal the plastic wrap tightly around the container. Place the rock in the center of the plastic wrap, directly above the cup. Put the container in a sunny spot.

4 After two days, observe the plastic wrap and the water in the container. Remove the plastic wrap and observe the plastic cup. Record your observations.

Wrap It Up!

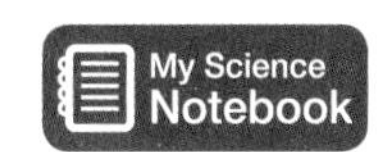

1. **Infer** What processes caused the water to move from the container into the cup?
2. **Compare and Contrast** Describe the water in the cup and the water in the container.
3. **Draw Conclusions** What can you conclude about how solar energy can be used to make water cleaner?

Obtain and Combine Information

You've seen many different ways that people affect the environment. You've also seen how people are working to protect the environment. Now you'll research and present information to answer this question: In what ways do individual communities use science ideas to protect Earth's resources and the environment?

1. **Carry out your research.** 

   Pick a resource that interests you, such as energy use, water quality, or wildlife. Identify a local group that is working to protect this resource, such as a city government or a community organization. If possible, interview the people who are doing the work. Collect information from sources such as newspaper articles, brochures, and websites. Record the source of all the information you obtain.

2. **Analyze and organize your research.**
   Study the data you have collected. Identify the important science ideas that relate to your selected natural resources. How have people used these ideas to help protect the environment? Has their work been successful?

   Think about how to organize your data. You might make a time line, or divide the information into categories, such as "Problems" and "Solutions."

PE 5-ESS3-1. Obtain and combine information about ways individual communities use science ideas to protect the Earth's resources and environment.

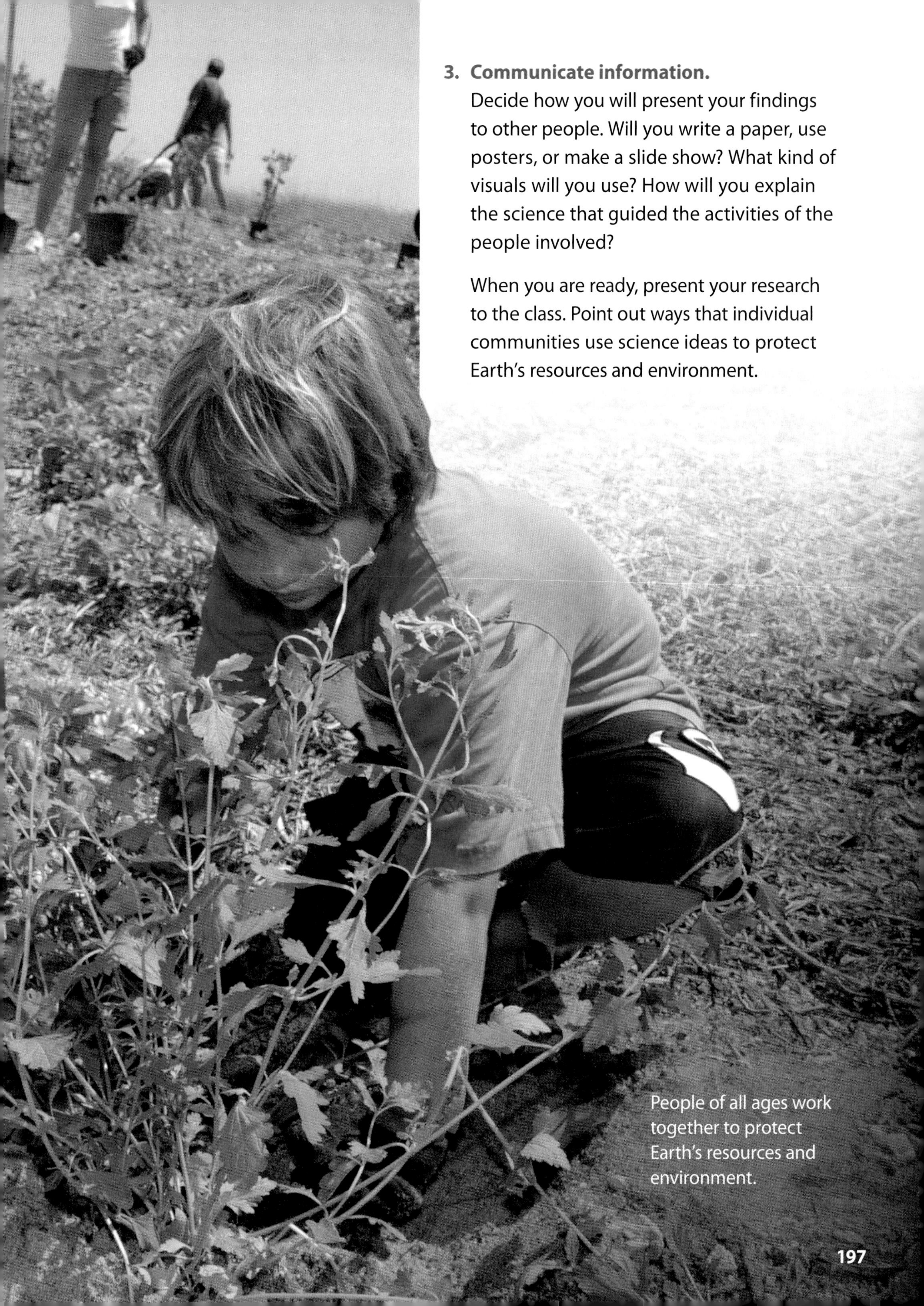

3. **Communicate information.**
   Decide how you will present your findings to other people. Will you write a paper, use posters, or make a slide show? What kind of visuals will you use? How will you explain the science that guided the activities of the people involved?

   When you are ready, present your research to the class. Point out ways that individual communities use science ideas to protect Earth's resources and environment.

People of all ages work together to protect Earth's resources and environment.

Gravity on Earth

Have you ever heard the saying, "What goes up must come down"? Everything falls downward, or toward Earth's center, because of gravity. **Gravity** is a force that pulls one object toward another object. Earth's gravity pulls all objects toward the planet's center.

One factor that affects the strength of gravity's pull is the distance between the two objects pulling on each other. Earth's pull on objects farther from the surface is less than it is on objects closer to the surface.

A skydiver falls to Earth because of gravity.

No matter where around Earth a skydiver is falling, gravity pulls the skydiver "down," or toward Earth's center.

DCI PS2.B: Types of Interactions. The gravitational force of Earth acting on an object near Earth's surface pulls that object toward the planet's center. (5-PS2-1)
CCC Cause and Effect. Cause and effect relationships are routinely identified and used to explain change. (5-PS2-1)

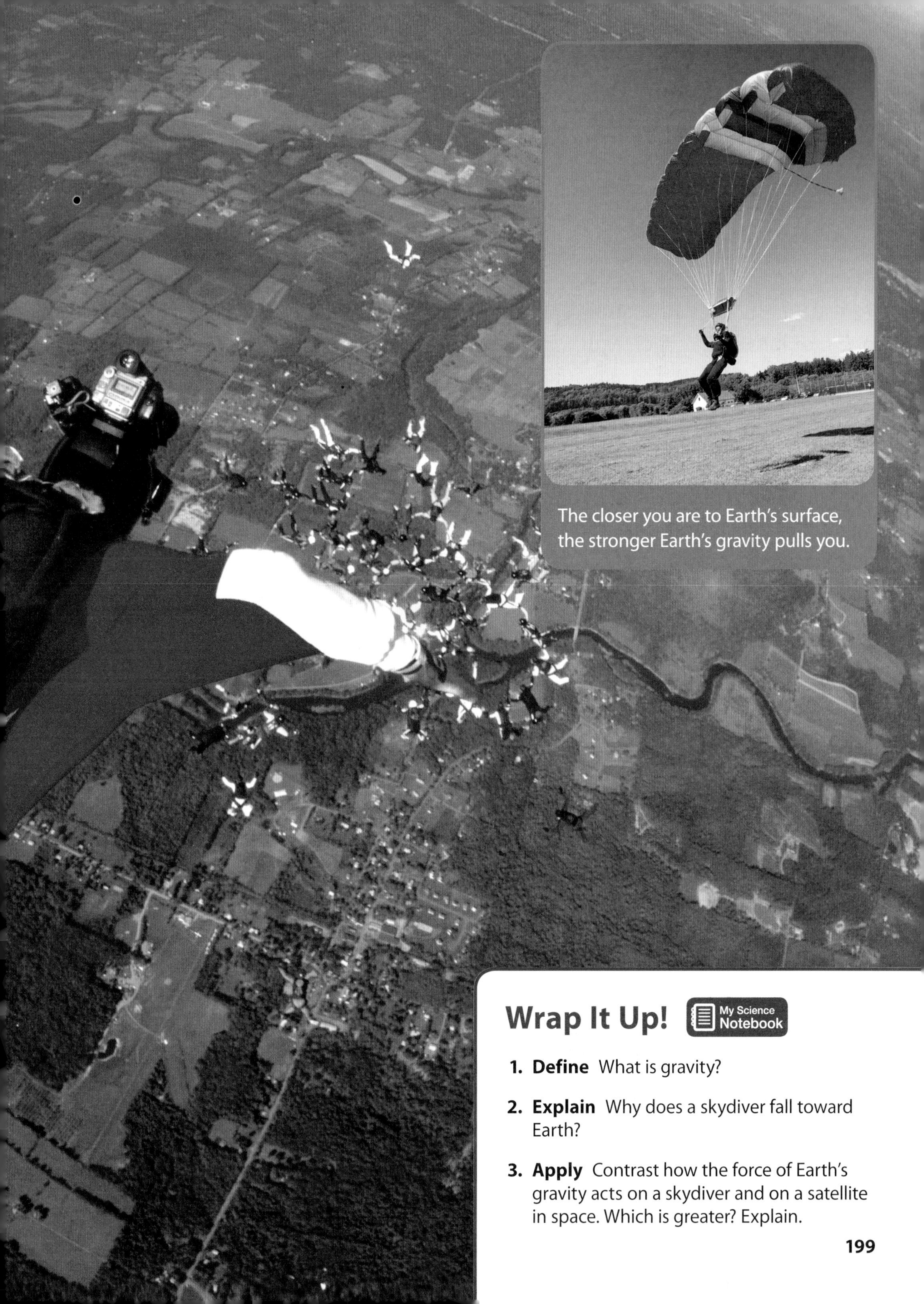

The closer you are to Earth's surface, the stronger Earth's gravity pulls you.

Wrap It Up! My Science Notebook

1. **Define** What is gravity?
2. **Explain** Why does a skydiver fall toward Earth?
3. **Apply** Contrast how the force of Earth's gravity acts on a skydiver and on a satellite in space. Which is greater? Explain.

Investigate

Gravity

How does gravity affect an object on Earth?

You've read about what happens when a skydiver jumps from a plane. In this investigation, you will observe what happens when objects fall near Earth's surface. Can you make objects move against gravity? Can you stop the fall of an object? Are objects always being pulled down toward Earth? You will use your observations to support an argument about how Earth's gravity affects objects.

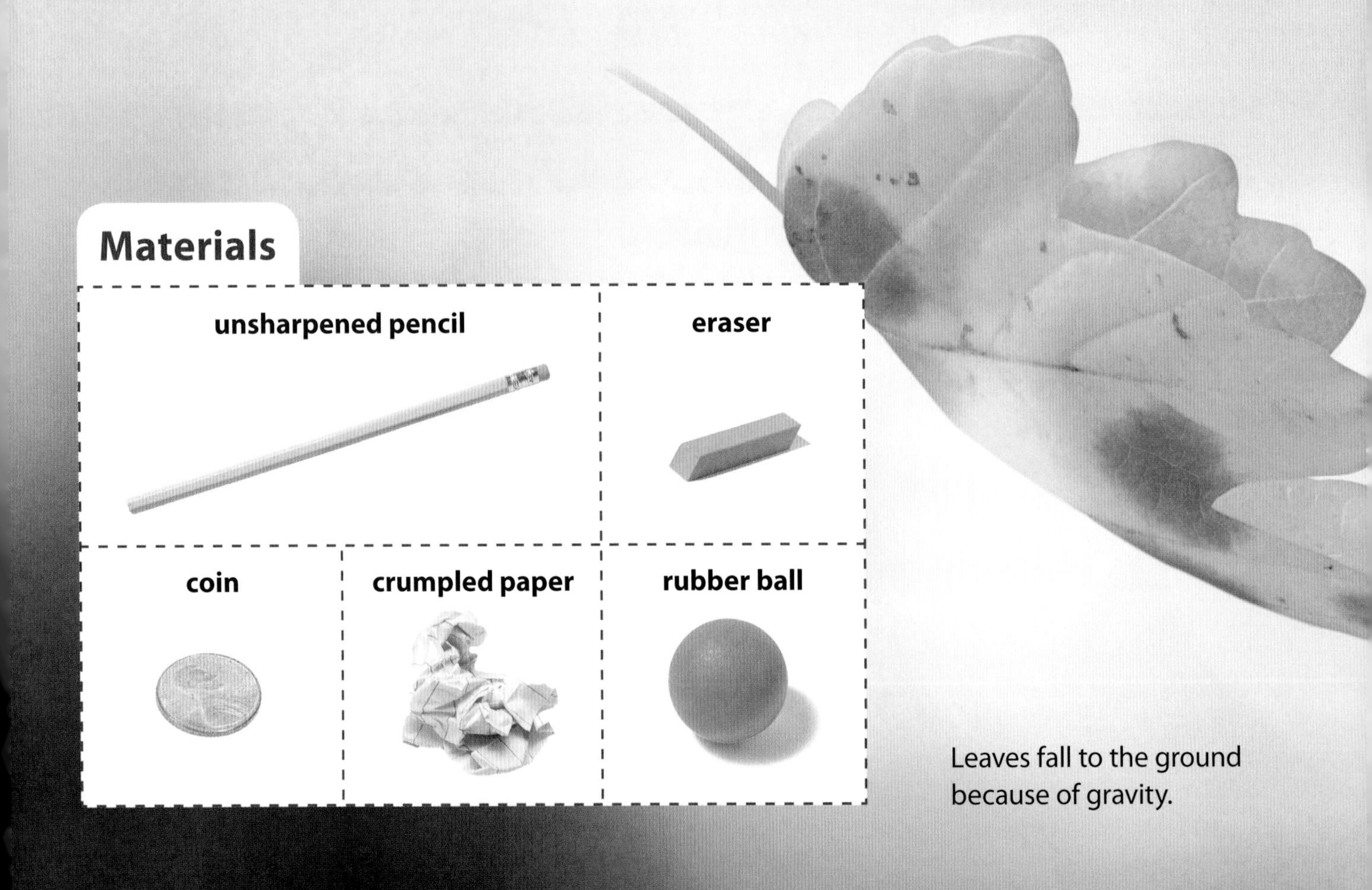

Leaves fall to the ground because of gravity.

PE 5-PS2-1. Support an argument that the gravitational force exerted by Earth on objects is directed down.

1 Observe each object, including its shape, size, weight, and what it is made of. Predict how Earth's gravitational force will affect each object as you drop it on a table, as you gently push it off the edge of a desk or table, and as you gently toss it in the air. Record your predictions in your science notebook.

2 Stand by a desk or table. Take turns with a partner. Drop one object from about shoulder height onto the desk or table. Record your observations. Repeat this for every object.

3 Gently push each object off of the edge of the desk or table. Record your observations in your chart. Repeat this for each object. Then compare your predictions with your observations.

4 Gently toss one object slightly up in the air. Do not toss it in your partner's direction. Record your observations. Repeat this for every object.

Wrap It Up!

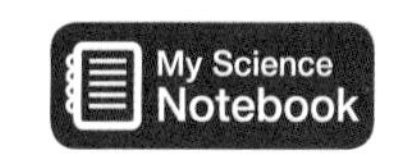

1. **Compare** Did your predictions support your results? Why do you think they were the same or different?
2. **Support an Argument** Use evidence from your investigation to support an argument that the force of Earth's gravity on an object is directed down.

Earth, Sun, and Moon

Two of Earth's close neighbors in space—the moon and the sun—form a system with Earth. They move independently and as a system. All three **revolve,** or move around other objects. Earth revolves around the sun, and—at the same time—the moon revolves around Earth. The sun, Earth, and moon together revolve around the center of our galaxy, the Milky Way.

The gravitational force between the sun and Earth keeps Earth in motion around the sun.

DCI PS2.B: **Types of Interactions.** The gravitational force of Earth acting on an object near Earth's surface pulls that object toward the planet's center. (5-PS2-1)

The motion of one object around another is called **revolution.** The sun, Earth, and moon revolve as the result of **gravitational force.** This force pulls objects toward each other. The gravitational force between Earth and the less-massive moon keeps the moon moving around Earth. The gravitational force between Earth and the sun keeps Earth revolving around this massive star.

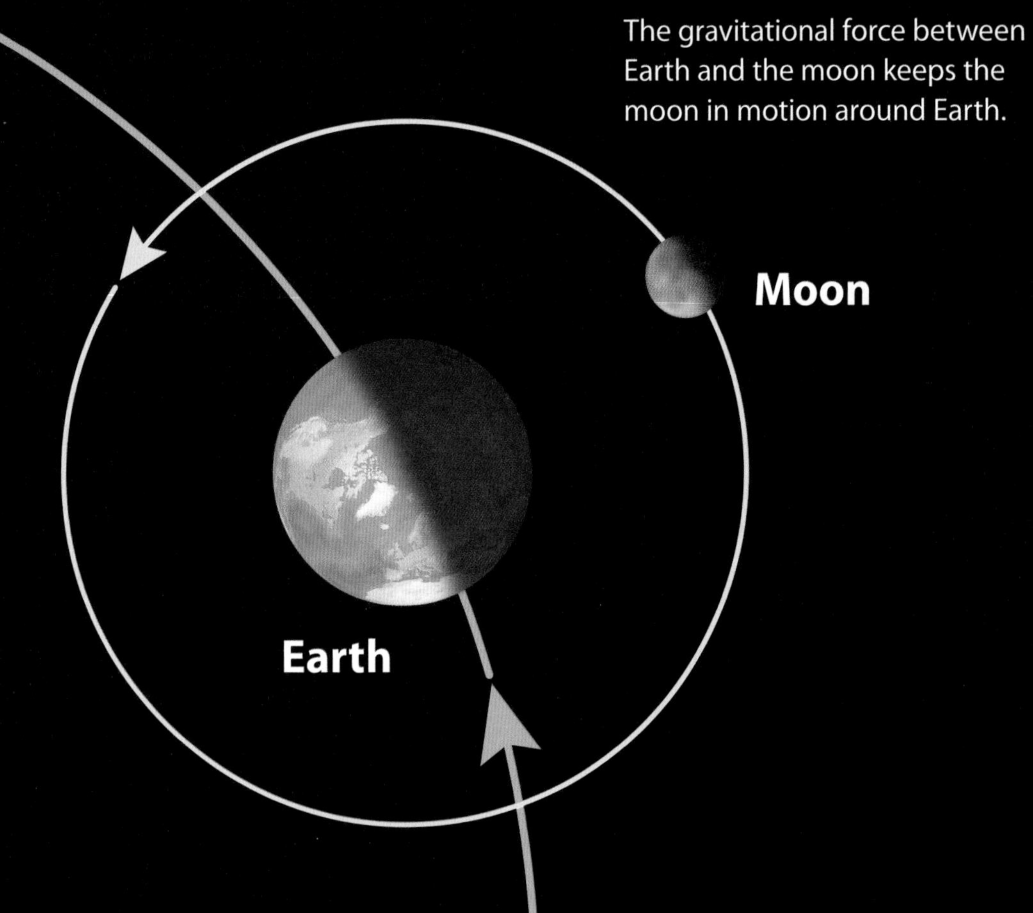

The gravitational force between Earth and the moon keeps the moon in motion around Earth.

This diagram is not drawn to scale. The sun's diameter is more than 109 times larger than the diameter of Earth. The actual distances between the moon, Earth, and the sun are far greater than shown.

Wrap It Up! 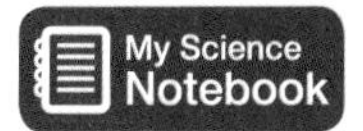

1. **Define** What is revolution?
2. **Explain** Tell why and how Earth, the moon, and the sun revolve.
3. **Infer** Gravitational force is related to mass. Infer which has a greater gravitational pull on Earth—the sun or the moon. Explain.

Citizen Science

Map the Moon

Expert Identifications

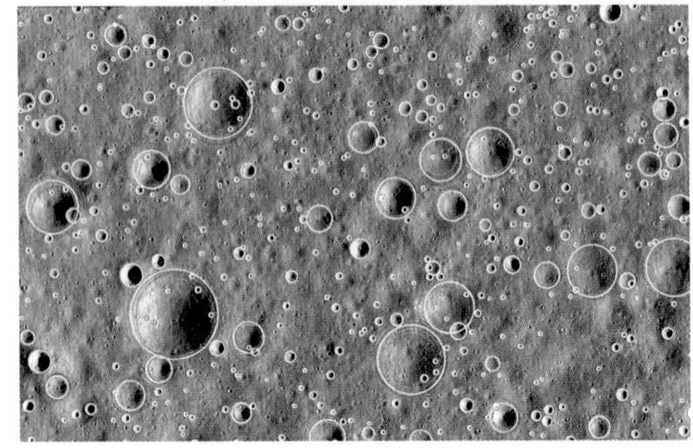

Volunteer Identifications

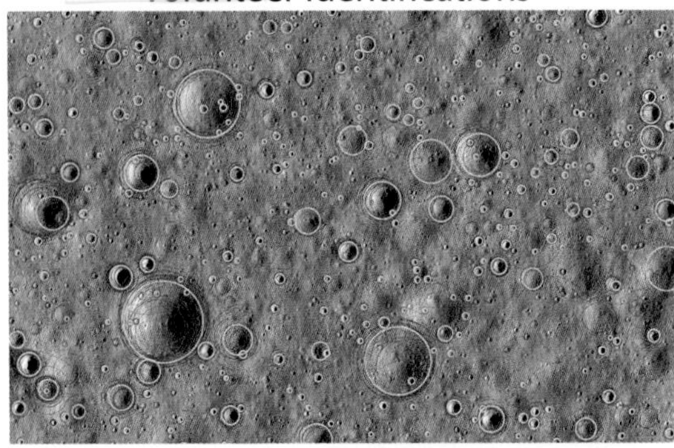

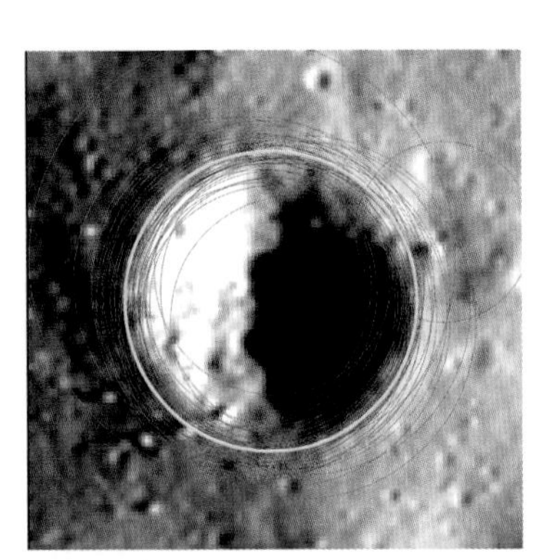

Citizen scientists collect data to help advance our knowledge of the moon's surface. Even expert scientists can use your help!

To safely travel to the moon, scientists need to map the major features, such as craters.

DCI PS2.B: Types of Interactions. The gravitational force of Earth acting on an object near Earth's surface pulls that object toward the planet's center. (5-PS2-1)
CCC Cause and Effect. Cause and effect relationships are routinely identified and used to explain change. (5-PS2-1)
NS Science Addresses Questions About the Natural and Material World. Science findings are limited to questions that can be answered with empirical evidence. (5-ESS3-1)

You can help scientists map our nearest rocky neighbor, the moon.

What Is Citizen Science?

Can you imagine that one day you or your children might take a quick vacation getaway to the moon? That day might not be too far off.

Like Earth's gravity, the moon's gravity causes objects in space to crash into its surface. But on the moon there is no weather, flowing water, or other ways to wear down craters.

So before we can do much moon travel safely, scientists need detailed maps of the moon. Scientists need your help! You can help scientists map our nearest rocky neighbor, the moon.

There are many more craters and rugged features to locate and map than any team of scientists can do alone. But there are plenty of people who like to observe the moon. Many are willing to take close-up looks at photos that scientists take and make sense out of them. By identifying craters and their size, they can help scientists make the detailed maps that are needed.

Citizen scientists are ordinary people who help gather data for real scientific studies. Many citizen scientists are using an online database called Moon Mappers to help find and map craters on the moon. A database stores data collected from many sources. The data is organized in a way that makes it easy for people to find and use it.

In Moon Mappers, your class will look at photographs of the moon's surface. You will need to practice identifying craters and using computer tools to place a circle around as many craters as you can see in each photo. You will have to make the circle the size of the crater and try to position the circle just right. But the practice is fun. Then your information gets added to the database. And in Moon Mappers, you can even listen to informative podcasts as you search for and circle each crater you find. There is a lot to learn.

You can work with a partner as you search for craters to circle. You will need to follow Moon Mappers rules for recording observations. And don't forget—you are doing real science as you help map the moon. Maybe one day you or your children will visit a crater you found on Moon Mappers!

Wrap It Up!

1. **Describe** What was the most difficult part of the Moon Mappers project? What did you like about the project?
2. **Explain** How did technology, such as computers, help you collect and share data?
3. **Compare and contrast** Describe the different photos you analyzed. How were the photos alike and different?

STEM

SCIENCE
TECHNOLOGY
ENGINEERING
MATH

SPACE STATION PROJECT

Toys on Earth and in Space

When you jump rope, you swing the rope around, jump over it, and gravity brings you back down to Earth. When you spin a top on a table, it will spin for a while and then topple over. Or it might spin right off the table and fall to the floor, then stop. When you use toys on Earth, they are affected by Earth's gravity. But in space, things are very different!

Think of a drop tower amusement park ride. You are at the top, and suddenly the ride is released. You drop and experience free-fall. You feel weightless, even though Earth's gravity is still affecting you and the ride. That is what is happening on the International Space Station. It is in free-fall, but it is also moving forward in an orbit around Earth. People feel weightless. This is called microgravity.

On the International Space Station, astronauts have tested some toys to see how they work in microgravity. You will use some toys on Earth and predict how they would work in microgravity. Then you will see what actually happens when they are used on the space station. Then you will work on your own toy challenge.

PE 3-5-ETS1-3. Plan and carry out fair tests in which variables are controlled and failure points are considered to identify aspects of a model or prototype that can be improved.

The Challenge

You will see how some common toys work on Earth and in microgravity in space. Your challenge is to design, build, and test a toy roller coaster that can work on Earth. Then you will make changes so that it would work in space.

Toys have been tested on the space shuttles and on the International Space Station. Here, astronaut Jeffrey Hoffman and mission specialist Rhea Seddon demonstrate how a coil toy works in near-weightless conditions on the Space Shuttle Discovery.

STEM

SPACE STATION PROJECT
(continued)

1 Define the problem. 

Think about the problem you are solving. The goals it must achieve are the criteria of the problem. You will know your design is successful if your model meets these criteria. Your roller coaster must:

- be made of materials your teacher supplies
- have one hill and one loop
- be able to keep a marble on track to complete the whole path of the roller coaster in conditions on Earth
- be analyzed and changed so the marble could complete the roller coaster path in microgravity

Constraints are things that limit your design. For example, cost can limit the materials you can use. Write the problem you need to solve in your notebook. List the criteria and constraints.

2 Find a solution.

Your teacher will give you directions for testing different toys to observe how they work on Earth. Then you will predict how each toy will work in space. You will view a video showing astronauts demonstrating toys in space.

After viewing the toys used in space, you will design your roller coaster. You will first design a roller coaster that will work on Earth. Consider these questions:

- How can you make sure the marble does not fall off of the roller coaster?
- When you release the marble, how will you make sure it can travel the complete path through the loop and over the hill of the roller coaster?

Sketch a design for your roller coaster. Discuss your teammates' designs. Choose a design you think will meet the criteria.

You are probably familiar with how toys are affected by Earth's gravity. How do you think these toys would work in conditions of microgravity on the International Space Station?

3 Test your solution.

Use your plan to build a roller coaster that will work at Earth's surface. Adjust your roller coaster to make it work. Make notes to show what you change. Test the roller coaster. Discuss the results of the tests with your team. Does the marble travel along the roller coaster without falling off? Does it complete the path of the roller coaster?

Now think about how your roller coaster would work in microgravity. How would you get the marble to move on the roller coaster? How could you make sure that the marble completes the whole path of the roller coaster? Are there places on the coaster where the marble would not stay in its path? What would happen to the marble as it traveled to the hill? Write and draw what you think would happen to the marble on the roller coaster. Then write and draw your thoughts for making the roller coaster work in microgravity.

4 Refine your solution.

Talk with your team about how you can change the roller coaster so that it would work in microgravity. Use your ideas to make changes. Then test the roller coaster again. Record your reasoning for making the changes. Compare your roller coaster design to other designs in the class. How do your results compare to the findings of other groups?

Present your roller coaster to the class. Describe the results of your tests. Explain how you had to change the design of the roller coaster so that it would work in microgravity.

Think about the results of your tests and feedback from the class. How could you improve your design? Record your ideas in your notebook.

Sally Ride in Space

The Challenger rockets into space on June 18, 1983. Sally Ride becomes the first American woman in space.

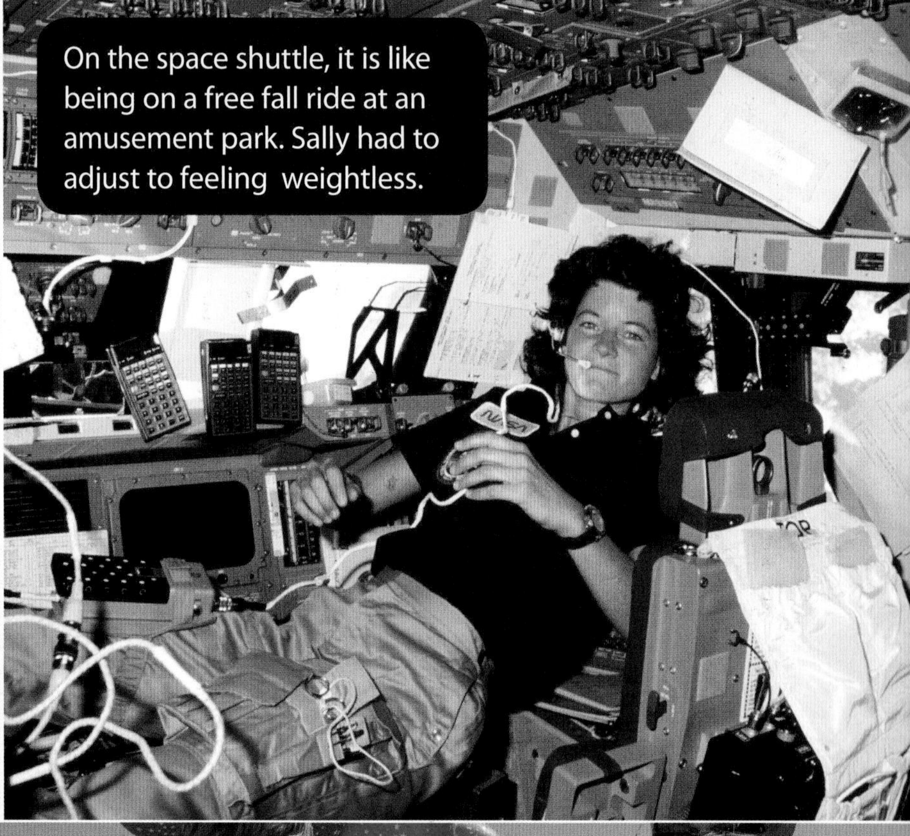

On the space shuttle, it is like being on a free fall ride at an amusement park. Sally had to adjust to feeling weightless.

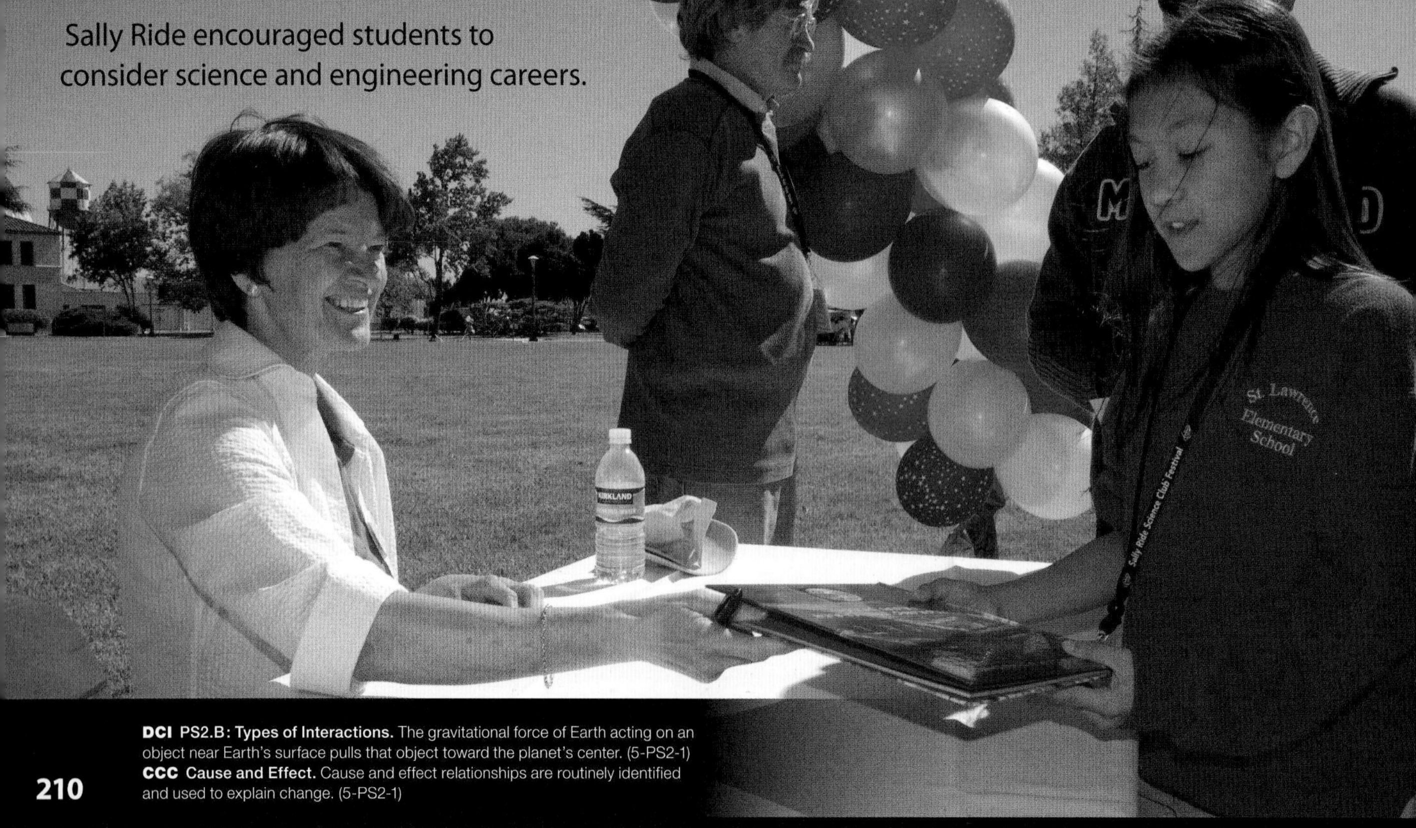

Sally Ride encouraged students to consider science and engineering careers.

DCI PS2.B: Types of Interactions. The gravitational force of Earth acting on an object near Earth's surface pulls that object toward the planet's center. (5-PS2-1)
CCC Cause and Effect. Cause and effect relationships are routinely identified and used to explain change. (5-PS2-1)

Sally Ride spent time encouraging young people, especially girls, to study math, science, engineering, and technology.

On June 18, 1983, Dr. Sally Ride became the first American woman to fly into space. Sally earned her doctorate in physics from Stanford University. She had always been interested in science, especially space and the space program. Sally was accepted into NASA's astronaut program in 1978. Her training to become an astronaut included a great deal of studying. She had to learn how each part of the space shuttle worked. She also learned about radio communications and navigation while flying in a jet. It was not Sally's job to fly the space shuttle. But all astronauts have to learn how to do the job in case of an emergency. Sally also had to take survival training. This included parachuting out of planes and exercising to have a high level of strength and fitness.

As if all of that is not exciting enough, Sally also had to learn how to deal with feeling weightless, as she would be in the microgravity conditions of space. One way to experience weightlessness on Earth is to send a plane into a high-speed dive. Experiencing weightlessness on Earth can often make people nauseated and dizzy. When Sally trained in the appropriately named "Vomit Comet" (the plane NASA used to practice the high-speed dives), she did not get sick. She was able to practice putting on a space suit and to use shuttle equipment while floating in the cabin of the plane.

Sally officially became an astronaut in 1979. When she flew her mission in 1983, she was the mission specialist. As part of her job, she operated the robotic arm attached to the outside of the shuttle to place satellites in Earth's orbit. She also collected data about Earth's atmosphere. The data collection included using high-resolution cameras and radar. She flew on a second mission in 1984. In an interview, Sally said the two most fun things about being in space were weightlessness and the view of Earth.

Sally left NASA in 1987. She began teaching physics at the University of California in San Diego. She also wrote science books for children. Sally Ride spent time encouraging young people, especially girls, to study math, science, engineering, and technology. The company she founded, Sally Ride Science, continues to offer summer camps for middle-school and high-school students that focus on these subjects. The camp activities include hands-on projects and workshops that provide stories from women working in these fields. Although Dr. Sally Ride died in 2012, her legacy lives on to inspire people to live their dreams.

Wrap It Up!

1. **Explain** Why do you think it was important for Sally Ride to practice being weightless on Earth before going into space?
2. **Relate** Tell about someone who inspires you to study science or engineering or to live your dreams by studying something else that you love to do.

Our Star—the Sun

A **star** is a ball of hot gases that gives off light and other types of energy. Stars range greatly in size and in the amounts of light and energy they give off. Even though the sun appears to be quite large to us on Earth, it is really a medium-sized star.

The sun appears larger and brighter than other stars in the sky because it is the closest star to Earth. It is so close, in fact, that its brightness blocks out the light of larger and brighter stars that are farther away. That's why you can't see other stars during daylight hours.

The sun is closer to Earth than the other stars. That makes its light look brighter than stars that are farther away.

DCI ESS1.A: The Universe and its Stars. The sun is a star that appears larger and brighter than other stars because it is closer. Stars range greatly in their distance from Earth. (5-ESS1-1)
CCC Scale, Proportion, and Quantity. Natural objects exist from the very small to the immensely large. (5-ESS1-1), (5-PS1-1)

Even stars that are bigger and brighter than the sun are visible only at night.

SCIENCE in a SNAP

Size and Distance

1 Put an object on your desk or other surface. Measure the object's length, or its distance across the middle if it is round. Record your measurement.

2 Move at least 2 m away from the object. Hold up your ruler to measure the apparent size of the object from your new position. Record your measurement.

? **How did the size of the object appear to differ when you viewed it from different distances?**

Wrap It Up!

1. **Describe** What is a star?
2. **Explain** Why does the sun seem to be brighter than other stars in the sky?
3. **Explain** Why does the sun appear to be so much larger than other stars in the sky?

Investigate

Apparent Brightness

? How does a light's brightness appear to change with distance?

You know the sun appears larger and brighter than other stars in the sky. That is because it is the closest star to Earth. In this investigation, you'll observe how distance affects brightness. Can you make a light from a flashlight appear brighter? Can you make it appear less bright? You will use your observations to support an argument about the cause of differences in the apparent brightness of the sun compared to other stars.

Each star's size and distance from Earth affects how bright it appears to us.

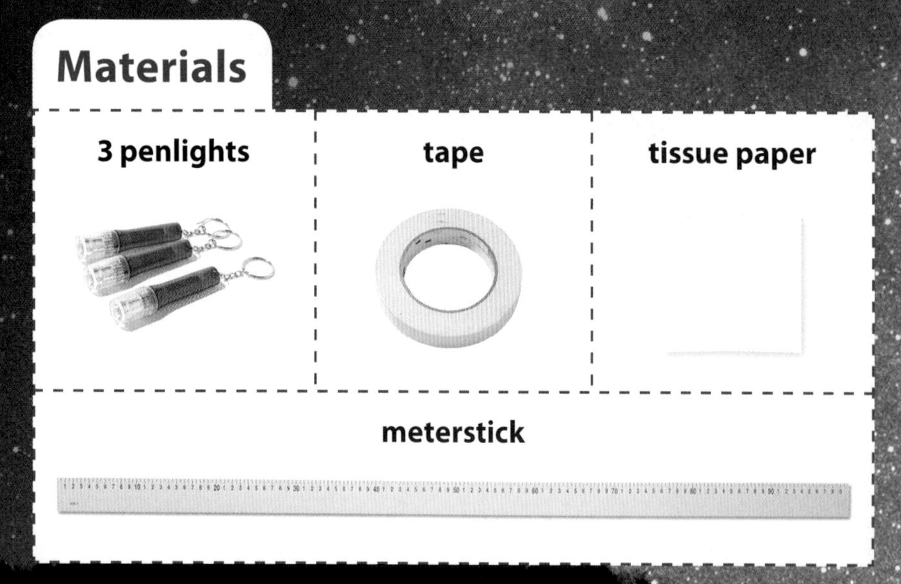

PE 5-ESS1-1. Support an argument that differences in the apparent brightness of the sun compared to other stars is due to their relative distances from Earth.

My Science Notebook

1 Make a chart for your data. Use the tape and a pencil to label the penlights as **A, B,** and **C.**

Trial	Penlight	Distance from Observer	Observed Brightness
	A	2m	
1	B	2m	
	C	4m	
	A	4m	
	B	2m	
2	C	6m	

2 Use the meterstick to measure three distances—2 m, 4 m, and 6 m—from a wall, desk, or table. Mark and label each distance with a piece of tape.

3 **TRIAL 1:** Have each person cover a penlight with a piece of tissue paper and stand at the 4-m mark. Tell them to turn on the penlights. Observe and record brightness using these words: *very bright, bright,* or *dim.*

4 **TRIAL 2:** Repeat step 3 with one person at the 4-m mark, one at the 2-m mark, and one at the 6-m mark. The person at the 6-m mark should be closest to you. Rank and record the brightness of each light as *very bright, bright,* or *dim.*

Wrap It Up!

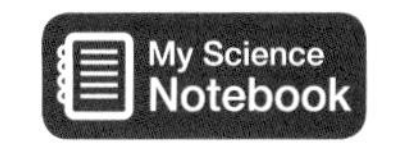

1. **Compare** Describe the brightness of the model stars in Trial 1.
2. **Infer** Which of the model stars in Trial 2 could represent the sun? Explain.
3. **Support an Argument** Why can stars with the same brightness appear dimmer or brighter than they actually are? Use your observations to support your argument.

Day and Night

Think about what happens when you **rotate,** or spin around. As you turn, you face different directions and see different things. Places on Earth are like that. Earth does not stand still. It rotates on an imaginary line called an **axis** that runs through the North and South Poles.

Half of Earth always faces the sun and is lit by sunlight. There it is day. The other half of Earth faces away from the sun and is dark. There it is night. Earth makes one complete rotation approximately every 24 hours. That's why one complete day–night cycle lasts 24 hours.

In about 24 hours, most areas on the surface of Earth rotate into sunlight, then into darkness, and then back into sunlight.

DCI ESS1.B: Earth and the Solar System. The orbits of Earth around the sun and of the moon around Earth, together with the rotation of Earth about an axis between its North and South poles, cause observable patterns. These include day and night; daily changes in the length and direction of shadows; and different positions of the sun, moon, and stars at different times of the day, month, and year. (5-ESS1-2)

CCC Patterns. Similarities and differences in patterns can be used to sort, classify, communicate and analyze simple rates of change for natural phenomena. (5-ESS1-2)

SCIENCE in a SNAP

Day and Night

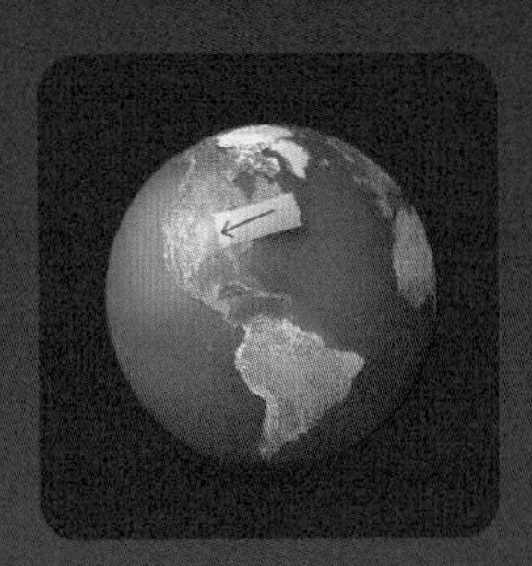

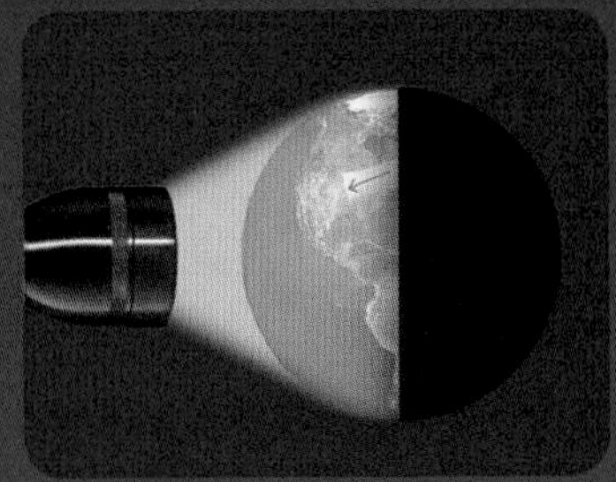

1 Mark your location on a globe with a piece of masking tape. Use a flashlight to represent the sun. Shine it on the globe so that it is day where you live.

2 Then rotate the globe so that it is night where you live.

? **What happened to the place on the globe where it used to be day? How often does this change happen?**

At this location, the sun will seem to move out of view as Earth rotates. But it is Earth's motion, not the sun's, that causes the change from day to night.

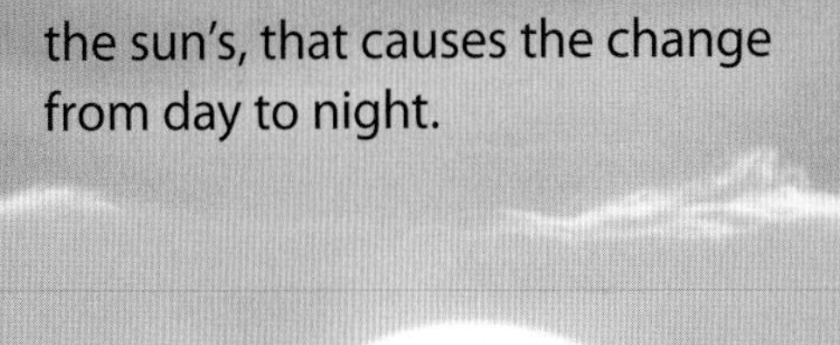

Wrap It Up!

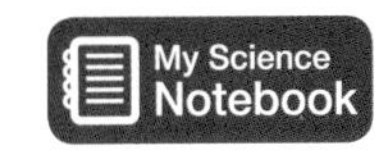

1. **Cause and Effect** How does Earth's rotation on its axis cause day and night?
2. **Apply** Which time period is closer to the scientific meaning of a day—the time period between sunrise and sunset or the 24-hour time period between one sunrise and the next sunrise?

Apparent Motion

The people on this ride are moving in a circle. When they look outward, the people standing still on the ground appear to be moving. But the bystanders have only **apparent motion,** or motion that *seems* to happen.

Like the turning ride, Earth's rotation causes some apparent movements. As observers look outward, they see the apparent motion of the sun across the sky from east to west. But it is Earth that is turning instead.

Like the sun and moon, many stars appear to rise above the horizon and move in an arch shape across the sky. In the Northern Hemisphere, stars high in the sky appear to move in a circle around the North Star.

DCI ESS1.B: Earth and the Solar System. The orbits of Earth around the sun and of the moon around Earth, together with the rotation of Earth about an axis between its North and South poles, cause observable patterns. These include day and night; daily changes in the length and direction of shadows; and different positions of the sun, moon, and stars at different times of the day, month, and year. (5-ESS1-2)

CCC Patterns. Similarities and differences in patterns can be used to sort, classify, communicate and analyze simple rates of change for natural phenomena. (5-ESS1-2)

From the point of view of the riders, objects on the ground appear to be moving.

SCIENCE in a SNAP

Observe Apparent Motion

1 Choose a place outdoors. Face south and draw a landmark that you observe. Label east and west. Observe where the sun appears. Draw the sun and record the time. CAUTION: Do not look directly at the sun.

2 Every two hours, return to exactly the same location. Repeat your observation. Describe how the sun appears to move in the daytime sky.

? **What causes this pattern?**

Wrap It Up!

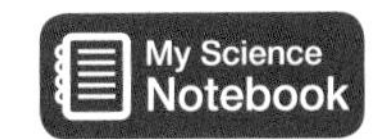

1. **Contrast** How are real and apparent motions different?
2. **Cause and Effect** Why do the sun, other stars, and the moon appear to change positions in the sky?
3. **Analyze** You are on the ground. You see a plane moving across the sky. Does the plane show real or apparent motion? Explain.

Investigate

Sunlight and Shadows

? How does a shadow in sunlight change during the day?

At dawn, your location on Earth's surface moves out of darkness and into sunlight. As Earth rotates, the sun appears to move across the sky. Shadows shorten and lengthen again. In this investigation, you can observe how the changing angle between the sun and your location on Earth's surface affects shadows caused by sunlight.

The location of the rider's shadow is determined by the position of the sun in the sky.

Materials

- clay
- poster board
- pencil
- colored pencil
- meterstick

PE 5.ESS1-2. Represent data in graphical display to reveal patterns of daily change in length and direction of shadow, day and night, and the seasonal appearance of some stars in the night sky.

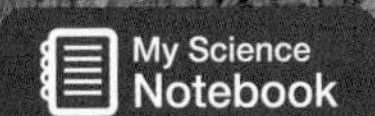

1 Put the poster board in a sunny place. Put a clay ball on the poster board. Push the end of a pencil into the clay.

2 Mark an ***X*** to show the direction of the sun. Record how high the sun looks in the sky. Trace the pencil's shadow. Write the date and time next to the shadow outline and the *X*.

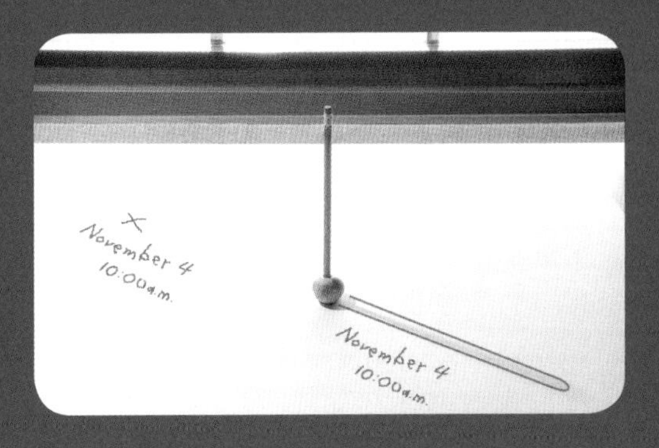

3 Use a meterstick to measure the shadow. Record the measurement and the time in your science notebook. Then repeat steps 2 and 3 at three more times during the day.

4 Use a colored pencil to draw where you predict the shadow will be in one hour. After one hour, repeat steps 2 and 3 one more time.

Explore on Your Own

What would you observe if you repeat the investigation in at least one other season? Plan and carry out your own investigation. Record your observations. Compare the results of your investigations.

Wrap It Up!

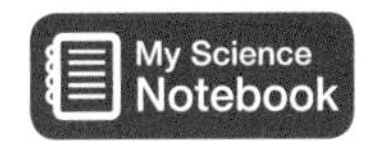

1. **Identify** What patterns in length and movement did you observe with the shadows?
2. **Explain** Did your results support your prediction? How is the sun's position related to the position and length of the shadows?
3. **Explain** How does Earth's rotation affect the appearance of a shadow caused by the sun?

Revolution and the Seasons

You can't feel it, but Earth moves. It rotates around its imaginary axis, which is tilted at an angle. Earth also revolves, or follows a regular path, around the sun. The path a revolving body follows is its **orbit.** Earth's tilt on its axis and its revolution around the sun cause observable patterns called seasons. Weather and the number of daylight hours at most places on Earth change with the seasons.

Earth's tilt on its axis also causes the seasons to be opposite in the Northern and Southern Hemispheres. For example, when it is summer in the Northern Hemisphere, it is winter in the Southern Hemisphere. Study the diagrams to see how Earth's changing position causes seasons. The captions describe seasonal changes in the Northern Hemisphere. Note that the sizes of Earth and the sun and the distance between them are not to scale in the drawings.

Earth's rotation and its revolution around the sun produce observable patterns.

These diagrams are not drawn to scale. The sun's diameter is many times larger than the diameter of Earth. The actual distance between Earth and the sun is far greater than shown.

DCI ESS1.B: Earth and the Solar System. The orbits of Earth around the sun and of the moon around Earth, together with the rotation of Earth about an axis between its North and South poles, cause observable patterns. These include day and night; daily changes in the length and direction of shadows; and different positions of the sun, moon, and stars at different times of the day, month, and year. (5-ESS1-2)
CCC Patterns. Similarities and differences in patterns can be used to sort, classify, communicate and analyze simple rates of change for natural phenomena. (5-ESS1-2)

Winter During winter the sun's energy strikes the Northern Hemisphere least directly. This causes fewer daylight hours and the weather to be colder.

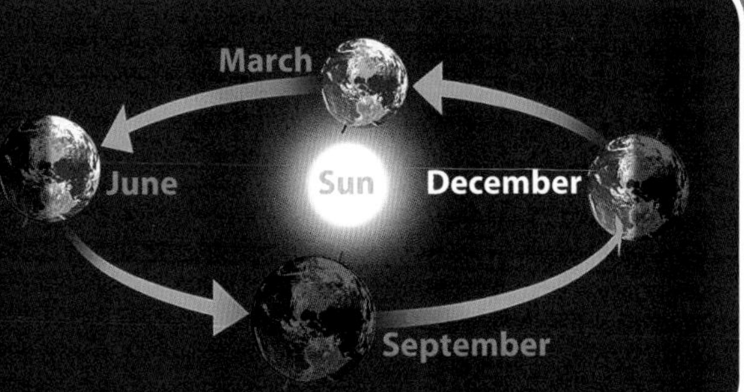

Spring In spring the sun's energy strikes Earth more directly than it did in winter. The weather warms and daylight hours increase.

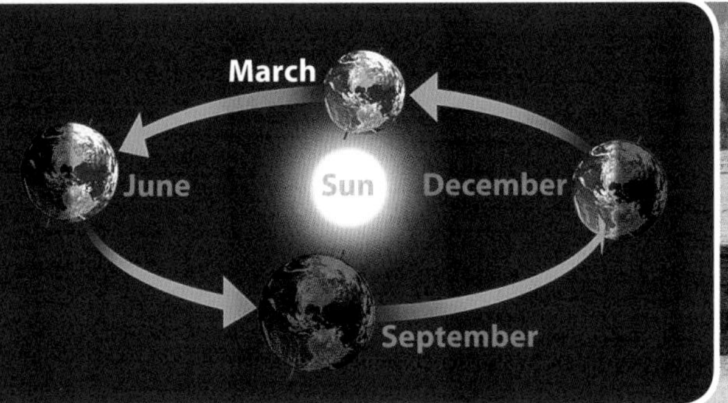

Summer Summer is the warmest season. Why? Summer days have more hours of sunlight. The sun's energy strikes parts of the Northern Hemisphere almost straight on.

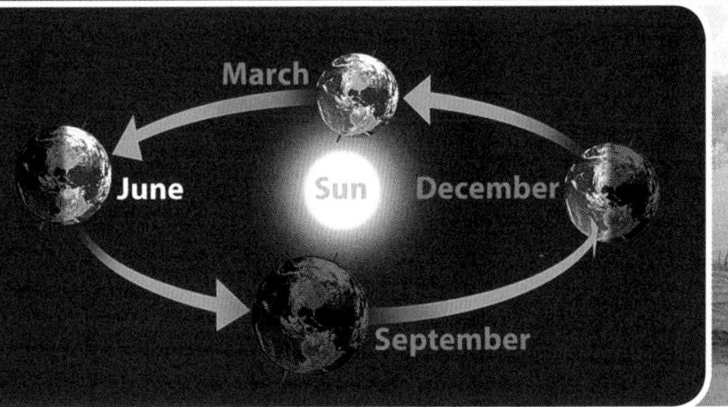

Fall Because of Earth's tilt, the sun's energy strikes less directly in the fall than in summer. The weather cools and hours of daylight decrease.

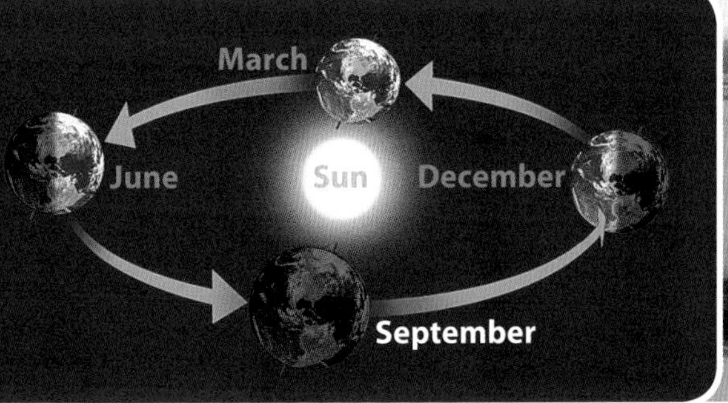

Wrap It Up!

1. **Recall** What causes seasons?
2. **Sequence** Beginning with winter, name the seasons in the correct order.
3. **Summarize** Describe how the number of hours of daylight changes with the seasons.

Investigate

Graph Hours of Daylight

What is the pattern of daylight hours throughout the year?

Depending on where you live, summer days can seem to go on forever! The number of hours of daylight at most places on Earth changes with the seasons. In this investigation, you'll plot data on a graph to reveal patterns of the average number of daylight hours in Chicago, Illinois, during a year.

Materials

graph paper

data table

Average Number of Daylight Hours per Month in Chicago, Illinois

Month	Average Number of Daylight Hours	Month	Average Number of Daylight Hours
January	296	July	460
February	308	August	428
March	370	September	374
April	400	October	344
May	451	November	296
June	455	December	286

PE 5-ESS1-2. Represent data in graphical displays to reveal patterns of daily changes in length and direction of shadows, day and night, and the seasonal appearance of some stars in the night sky.

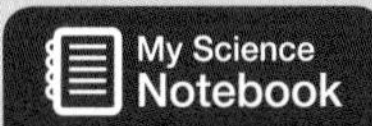

1 Make a graph. Label it as shown.

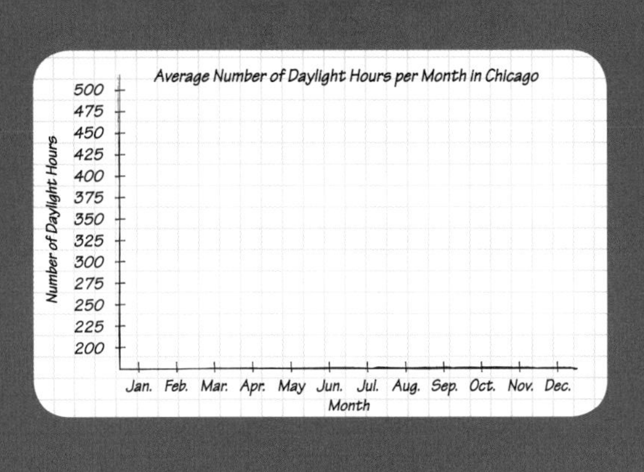

2 Plot the average number of daylight hours for January on your graph.

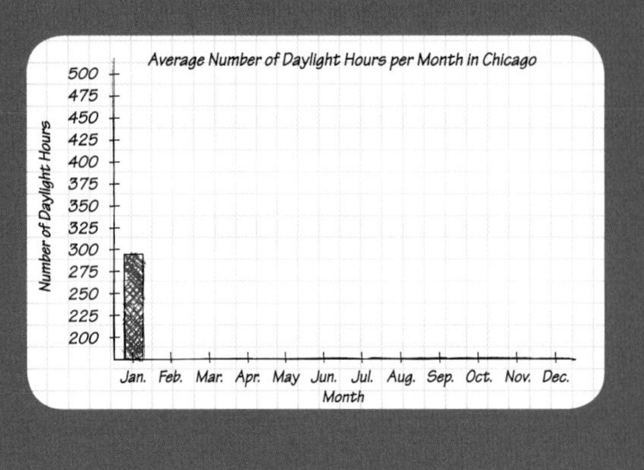

3 Then plot the average number of daylight hours for the other months.

4 Study your graph. Use it to answer the questions below.

The time of sunrise and sunset changes with the seasons.

Wrap It Up!

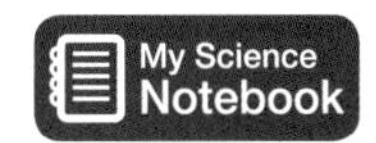

1. **Identify** Which months have the greatest and least average number of daylight hours?
2. **Summarize** Use your graphed data to describe patterns in the average number of daylight hours from month to month in Chicago.

Earth's Orbit and the Night Sky

Ancient people thought that some **constellations,** or star patterns, looked like animals or objects. They noticed that some constellations were visible only at certain times of the year. The position of the constellations in the sky changed during the time they were visible, too.

What caused these changes? Think about walking around the outside of your school building.

Fall In fall you can see Pegasus the flying horse.

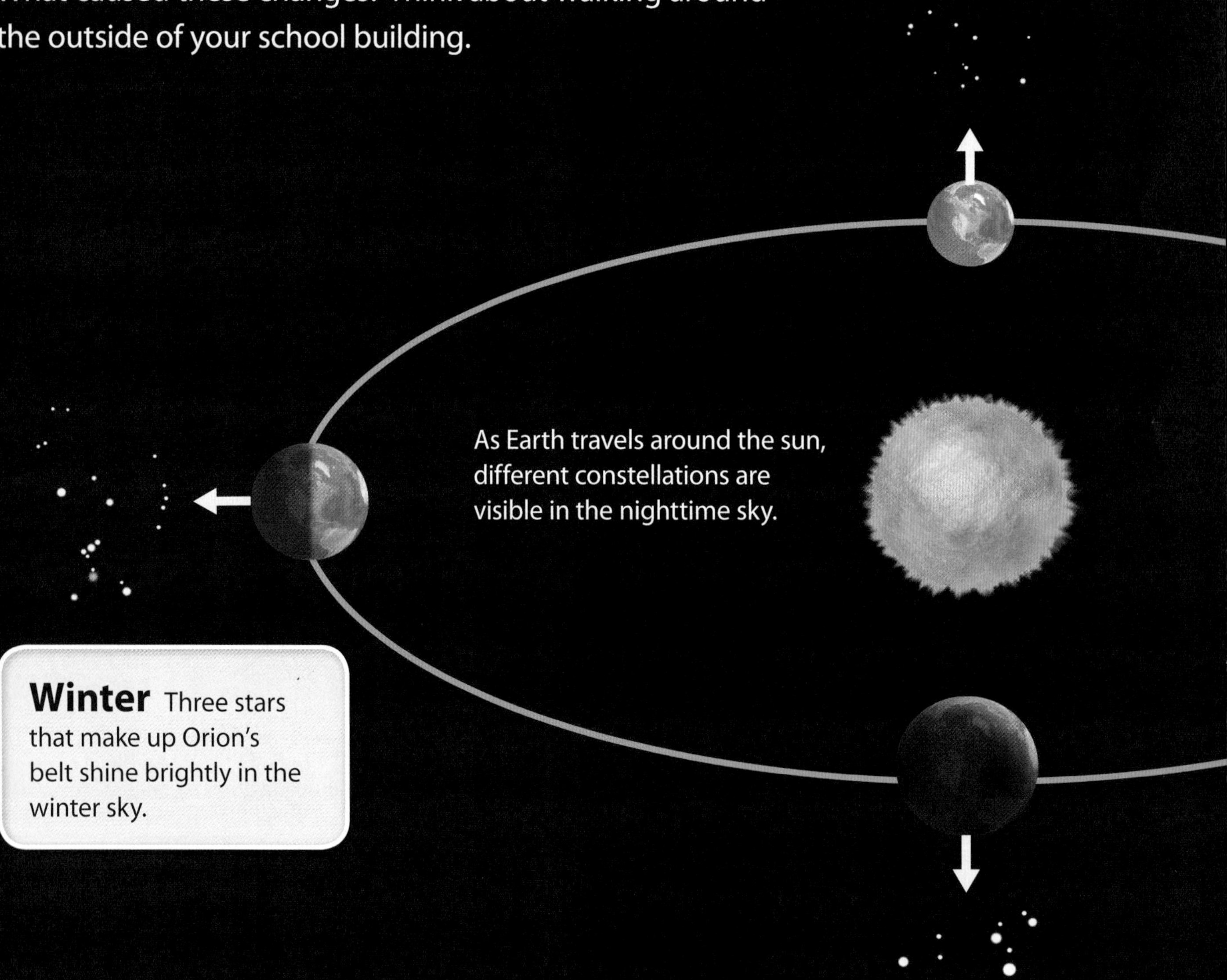

As Earth travels around the sun, different constellations are visible in the nighttime sky.

Winter Three stars that make up Orion's belt shine brightly in the winter sky.

Spring Leo the lion is a spring constellation.

DCI ESS1.B: Earth and the Solar System. The orbits of Earth around the sun and of the moon around Earth, together with the rotation of Earth about an axis between its North and South poles, cause observable patterns. These include day and night; daily changes in the length and direction of shadows; and different positions of the sun, moon, and stars at different times of the day, month, and year. (5-ESS1-2)

CCC Patterns. Similarities and differences in patterns can be used to sort, classify, communicate and analyze simple rates of change for natural phenomena. (5-ESS1-2)

As you look outward from the school, you can see only certain objects on certain sides of the school. Now think about Earth orbiting the sun. Earth's position changes throughout the year. Certain stars are visible only because we are looking outward into a different part of the night sky that is no longer blotted out by the sun's light.

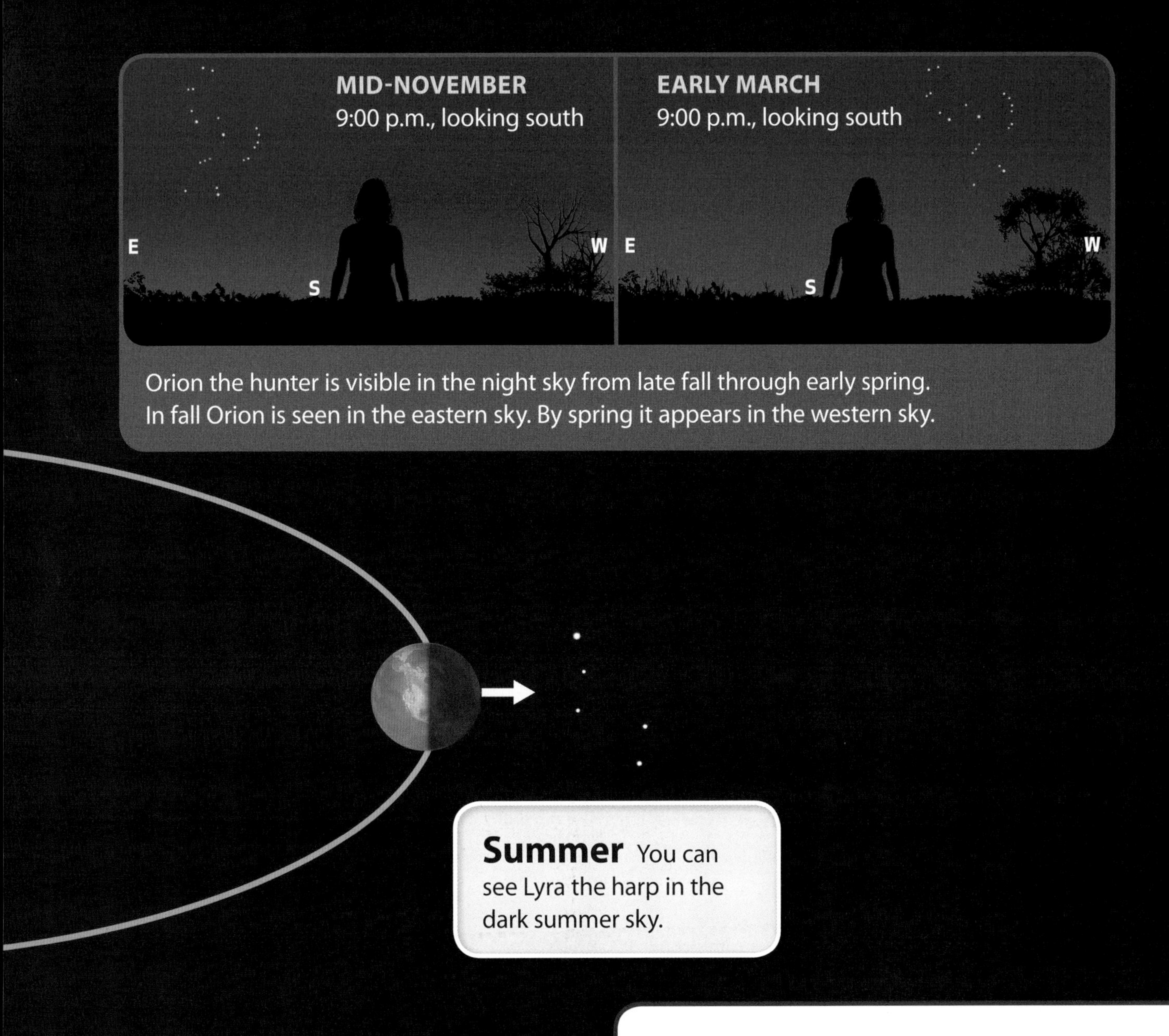

Orion the hunter is visible in the night sky from late fall through early spring. In fall Orion is seen in the eastern sky. By spring it appears in the western sky.

Summer You can see Lyra the harp in the dark summer sky.

This diagram is not drawn to scale. The sun's diameter is many times larger than the diameter of Earth. The actual distance between Earth and the sun is far greater than shown.

Wrap It Up!

1. **Explain** Why are most stars visible only at night?
2. **Cause and Effect** Why are some constellations only visible at cerrtain times of the year?

Represent Data

Remember the constellation Orion? It moves through the sky from late fall through early spring. In fall, if you look south at 9:00 p.m., you'll see Orion in the eastern sky. In December it is high overhead. By spring it is visible in the western sky.

You can apply what you know about Orion to other constellations. Study the diagram. Note the constellation listed for each season. These constellations would be high overhead in the night sky in the identified seasons. After you study the diagram, answer the questions.

1. **Identify.** 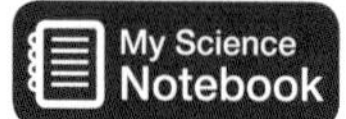

   Which constellation is high in the sky in spring? In summer? In fall? In winter?

2. **Represent data in graphical displays.**
   Choose one of the constellations shown. Draw a series of three illustrations. First, show where an observer would see the constellation early in the season. Second, show where an observer would see the constellation mid-season. Third, show where an observer would see the constellation late in the season. Then use your graphical displays to describe the pattern of the appearance of stars in the sky.

PE 5-ESS1-2. Represent data in graphical displays to reveal patterns of daily changes in length and direction of shadows, day and night, and the seasonal appearance of some stars in the night sky.

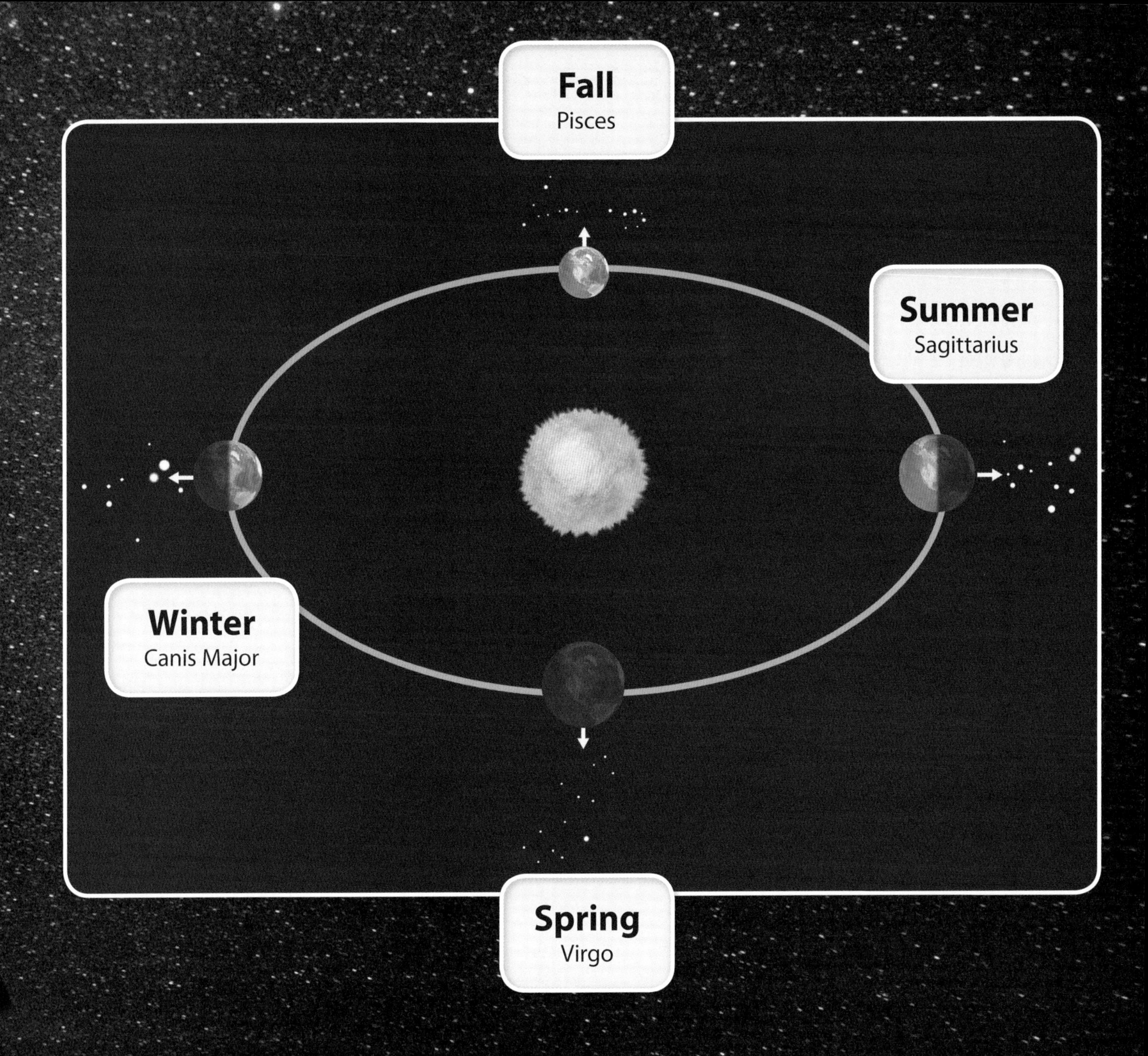

Which constellations are visible and where they appear in the sky depends on the time of year.

Moon Motions

Like Earth, the moon rotates, or spins, around an imaginary axis. The moon's orbit, however, is around Earth. As Earth completes its revolution around the sun once each year, the moon revolves around Earth about twelve times.

Motions of Earth and the moon result in observational patterns. Earth's rotation causes the moon to appear to rise toward the east, move across the sky, and set toward the west.

The apparent shape of the moon changes throughout the month in a repeating pattern.

DCI ESS1.B: Earth and the Solar System. The orbits of Earth around the sun and of the moon around Earth, together with the rotation of Earth about an axis between its North and South poles, cause observable patterns. These include day and night; daily changes in the length and direction of shadows; and different positions of the sun, moon, and stars at different times of the day, month, and year. (5-ESS1-2)

CCC Patterns. Similarities and differences in patterns can be used to sort, classify, communicate and analyze simple rates of change for natural phenomena. (5-ESS1-2)

The revolution of the moon around Earth causes the moon's shape to seem to change from day to day. The revolution of the moon around Earth is a key reason that the moon's apparent change in shape is a pattern that occurs again and again.

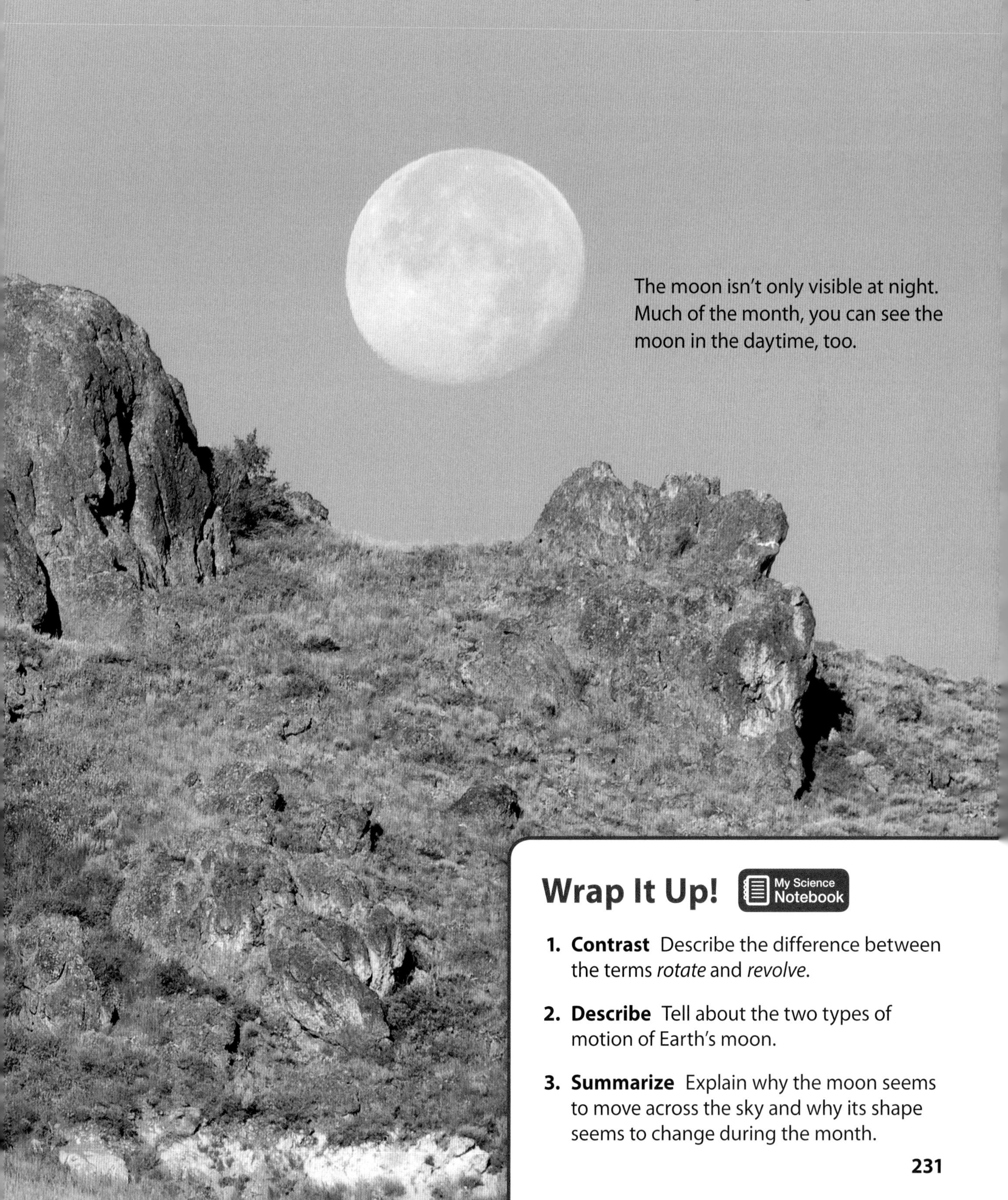

The moon isn't only visible at night. Much of the month, you can see the moon in the daytime, too.

Wrap It Up! My Science Notebook

1. **Contrast** Describe the difference between the terms *rotate* and *revolve*.
2. **Describe** Tell about the two types of motion of Earth's moon.
3. **Summarize** Explain why the moon seems to move across the sky and why its shape seems to change during the month.

the moon and the order in which they occur. A **phase** is how the moon appears from Earth. It takes the moon about four weeks to go through the cycle of phases.

As you study the pictures, you'll see that some days the moon looks like a full circle. Other days it looks like part of a circle. But the moon doesn't change shape. It is always shaped like a ball. What changes is how much of the moon's lighted half you can

see from Earth. Compare the phase pictures with the diagram on the next page. Like Earth, half of the moon always faces the sun. That side is lit. How much of this lit side is visible depends on where the moon is in its orbit around Earth.

DCI ESS1.B: Earth and the Solar System. The orbits of Earth around the sun and of the moon around Earth, together with the rotation of Earth about an axis between its North and South poles, cause observable patterns. These include day and night; daily changes in the length and direction of shadows; and different positions of the sun, moon, and stars at different times of the day, month, and year. (5-ESS1-2)

CCC Patterns. Similarities and differences in patterns can be used to sort, classify, communicate and analyze simple rates of change for natural phenomena. (5-ESS1-2)

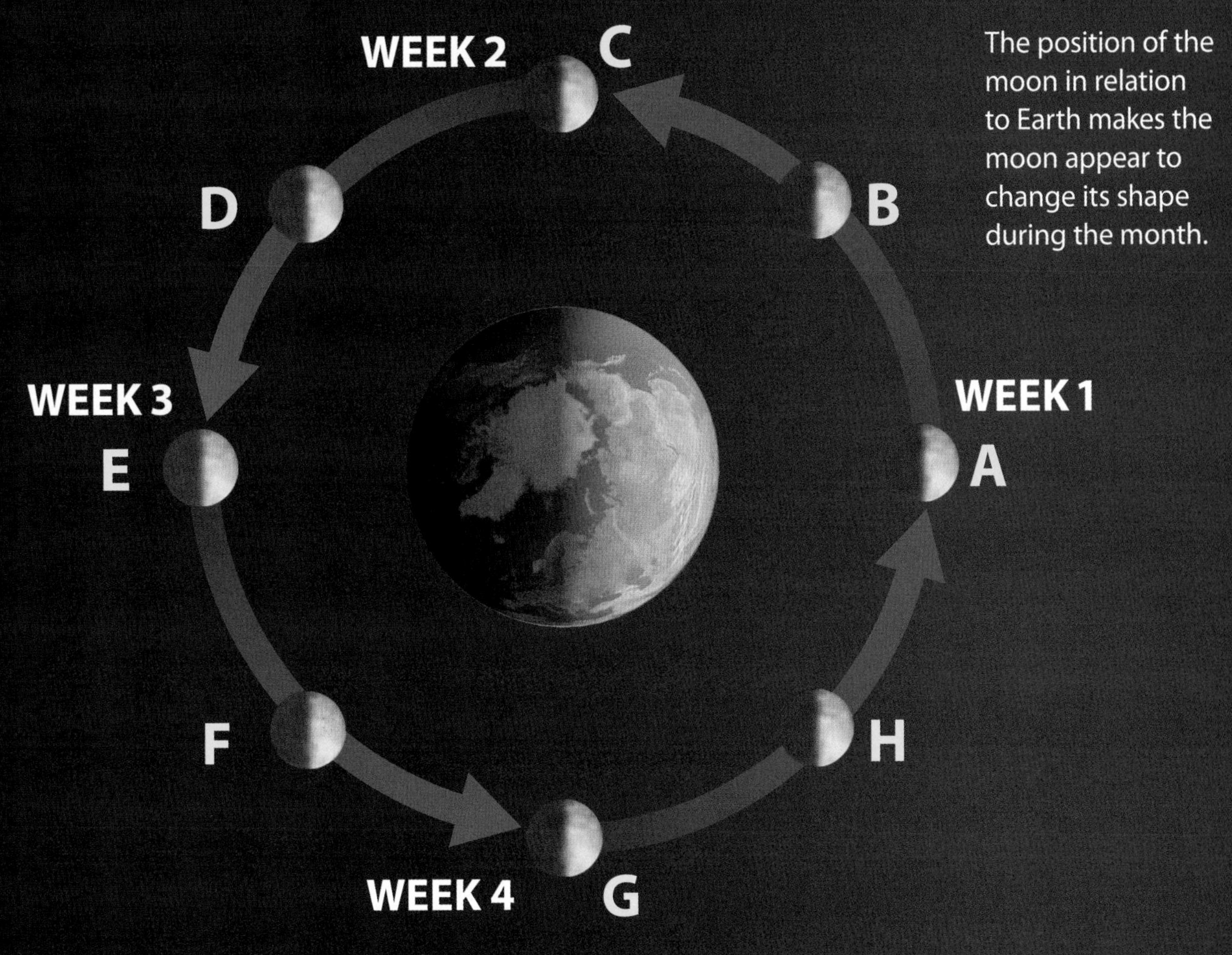

The position of the moon in relation to Earth makes the moon appear to change its shape during the month.

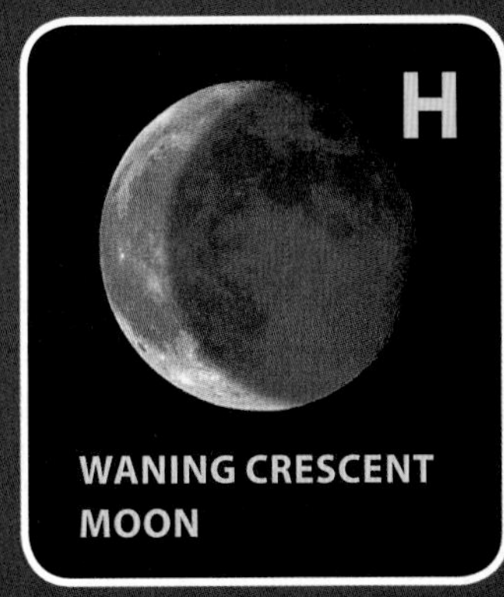

Earth's moon appears to change shape from day to day. These changes in shape are called moon phases.

Wrap It Up!

1. **Sequence** Name the eight phases of the moon, beginning with the new moon.
2. **Describe** Tell how the moon seems to change shape as it goes from new moon to full moon, and from full moon back to new moon.
3. **Summarize** Explain why the moon seems to change its shape during the month.

Investigate

Moon Phases

? How does the moon's orbit affect how it looks at different times of the month?

The moon's orbit around Earth results in different phases. In this investigation, you'll model how the moon goes through its phases.

The moon phases calendar shows how the appearance of the moon changes a little bit each day.

MOON PHASES CALENDAR

Sunday	Monday	Tuesday	Wednesday	Thursday	Friday	Saturday
1	2	3	4	5	6	7
8	9	10	11	12	13	14
15	16	17	18	19	20	21
22	23	24	25	26	27	28

Materials

- lamp
- ball
- craft stick
- meterstick

DCI ESS1.B: Earth and the Solar System. The orbits of Earth around the sun and of the moon around Earth, together with the rotation of Earth about an axis between its North and South poles, cause observable patterns. These include day and night; daily changes in the length and direction of shadows; and different positions of the sun, moon, and stars at different times of the day, month, and year. (5-ESS1-2)

CCC Patterns. Similarities and differences in patterns can be used to sort, classify, communicate and analyze simple rates of change for natural phenomena. (5-ESS1-2)

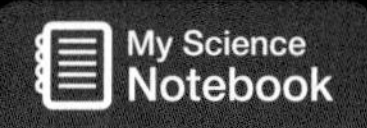

1 Push the craft stick into the foam ball. The ball is a model of the moon. Place the lamp at eye level. The lamp represents the sun. Stand 2 m from the "sun" and hold the "moon" in front of you. Your head represents Earth.

2 Have a partner stand near the lamp. Have a second partner stand behind you. Raise the moon just above your head.

3 Slowly turn your body in a circle to represent the moon revolving around Earth. Observe the light on the moon as you turn. Have your partners observe the moon, too. Record all of your observations in your science notebook.

4 Using the Moon Phases Calendar as a guide, make a plan with your partners to show the same pattern of moon phases using the moon model. Try your plan.

When viewed from the moon, Earth's appearance also changes!

Wrap It Up!

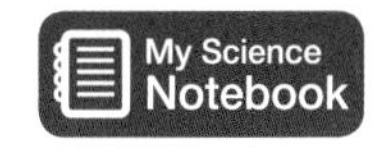

1. **Cause and Effect** In step 4, did you move in a clockwise or in a counterclockwise direction to produce the phases of the moon shown in the Moon Phases Calendar?
2. **Predict** If the moon is in the phase numbered 14, what phase will it be in the next day? In two weeks? In one month?
3. **Compare and Contrast** How is your model of the moon's motion and changing phases alike and different from the real Earth-Moon system?

Astrobiologist and Science Educator

How did life on Earth begin? Do living things exist elsewhere in our universe? These are just two of the many questions that astrobiologists like Brendan Mullan strive to answer. An astrobiologist is someone who studies the origin of life and how organisms can survive in extreme conditions. These scientists work to determine whether life could be supported in other places besides Earth.

When Brendan was in college, he was especially interested in the possibility of intelligent life on other worlds. He wanted to know if humans were alone in the universe. As Brendan studied, he started thinking about how to share his ideas with others. He wanted to find ways to make learning about science fun and interesting for students. He realized the importance of inspiring young people to pursue science careers.

NS Science Models, Laws, Mechanisms, and Theories Explain Natural Phenomena. Science explanations describe the mechanisms for natural events. (5-LS2-1)

Explorer

Brendan Mullan studied astronomy at Colgate University and Penn State University. He now designs astronomy courses for Penn State. He continues to share his knowledge and ideas with students of all ages through live presentations and shows. Through his efforts, he hopes to inspire the next generation of young scientists.

Brendan educates and inspires people of all ages with his astronomy presentations.

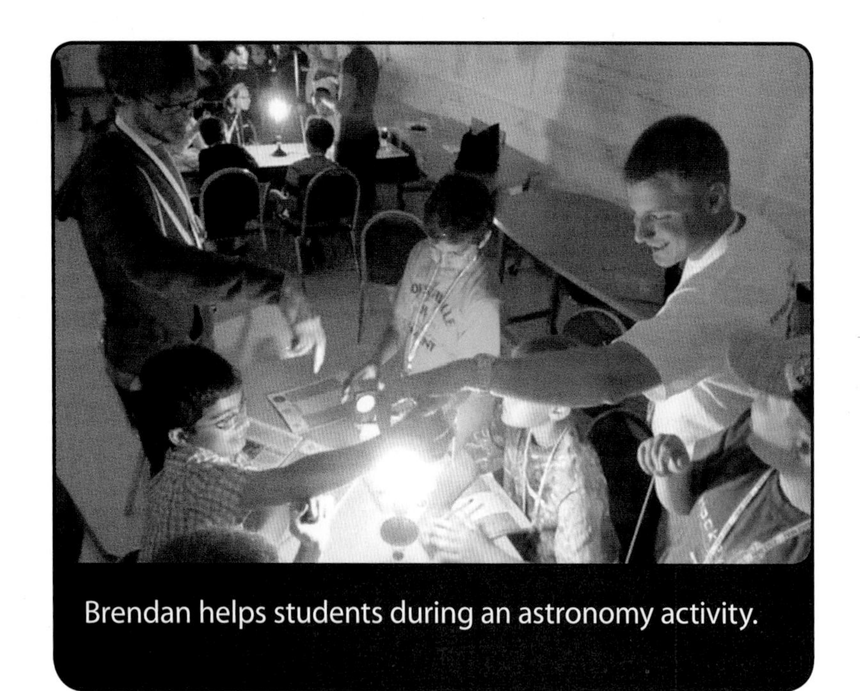
Brendan helps students during an astronomy activity.

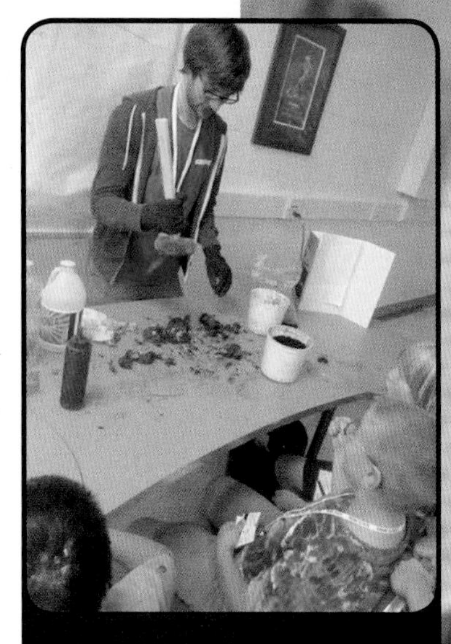
Students see a science activity up close.

Sharing scientific ideas with others is known as **science communication.** In 2012, Brendan won a national science communication competition called FameLab. He had three minutes to present an idea to the audience using only props that he could carry on stage.

Brendan's creative communication style is about more than entertainment. He has said, "If we can get people excited about astronomy, they'll use that passion to become not just astronomers, but the innovative biologists, chemists, physicists, and engineers that build the foundations of our modern society."

Brendan hopes that getting kids excited about science will encourage them to become the scientists and engineers of tomorrow.

Check In My Science Notebook

Congratulations! You have completed *Earth Science*. Now let's reflect on what you have learned. Here you'll find a checklist and some additional questions to help you assess your progress.

Page through your science notebook to find examples of different items. On a separate page in your science notebook, write your assessment of your work so far.

Read each item in this list. Ask yourself if you think you did a good job of it.

For each item, select the choice that is true for you: A. Yes **B.** Not Yet

- I defined and illustrated science vocabulary and main ideas.
- I labeled drawings. I wrote notes to explain ideas.
- I collected photos, news stories, and other objects.
- I used tables, charts, or graphs to record and analyze data.
- I included evidence for explanations and conclusions.
- I described how scientists and engineers answer questions and solve problems.
- I asked new questions of my own.
- I did something else. (Describe it.)

Reflect on Your Learning

1. What is one thing you learned that you can use in your everyday life? Explain.
2. Is there anything you still don't fully understand? What can you do to help yourself understand more clearly?

More to Explore

I discovered through my research that the Boiling River is caused by non-volcanic geothermal heating deep underground! The deeper you go, the hotter it gets. At the Boiling River, faults in Earth's crust allow hot water from deep in Earth to quickly flow to the surface! I was able to figure this out with the help of all the information I gathered and recorded in my notebook.

How have you thought and acted like a scientist or an engineer this year? How has your notebook helped you? What new information have you learned? Look back through your notebook, and think about some of the most important and interesting things you have written about. Discuss some ideas with your classmates. And remember to keep asking questions and finding answers as you continue exploring. Always be curious!

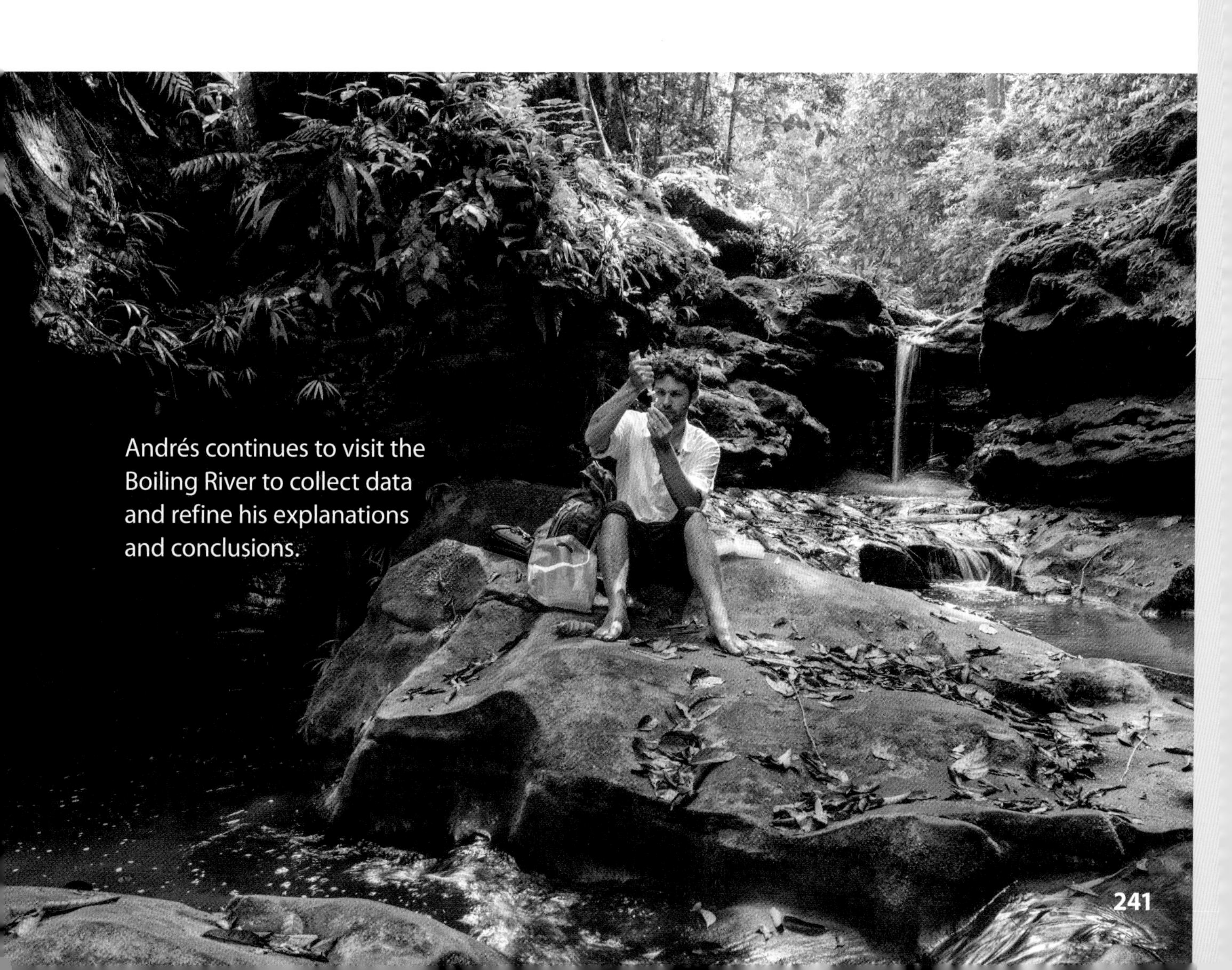

Andrés continues to visit the Boiling River to collect data and refine his explanations and conclusions.

Science Safety

When done safely, hands-on work can be one of the most fun and rewarding parts of studying science. Here are some important rules to follow whenever you are conducting a scientific investigation. Read these rules before you begin each investigation, and ask questions about anything you do not understand. Always wait for your teacher's permission before you begin an investigation or touch any equipment. If you get hurt, tell your teacher right away.

The Lab Space

- Be aware of the materials in your space and use them with care.
- Know where the first aid kit is kept.
- Know the location of the fire blanket.
- Sit only on lab chairs or stools, not lab tables.
- Keep your area neat.
- Immediately report an accident to an adult.

Lab Clothing

- Tie back loose or long hair.
- Do not let jewelry or clothing hang loosely.
- Shoes should cover the foot; do not wear sandals.
- Wear goggles, gloves, or an apron when necessary.

Chemical Safety

- Do not eat or drink anything in the science room.
- Never mix chemicals.
- Find out where to dispose of chemicals.
- Keep hands away from your eyes and mouth.
- Wash your hands when you are finished.

Glass Safety

- If you break any glassware, tell your teacher right away.
- Ask an adult to place broken glass in a container with a seal.

Fire and Electrical Safety

- Never reach over a heat source.
- Never touch a heated object without oven mitts.
- Stop-drop-roll if your clothes catch fire.
- Do not touch charged ends of batteries.
- Be aware of electrical cords.
- Keep all electrical equipment away from water.

Animal and Plant Safety

- Be responsible and appreciate the living things in your lab.
- Properly care for plants and animals.
- Only touch animals and plants with permission from your teacher.
- Wash hands after handling plant or animal material.

Cleanup

- Close all containers.
- Return materials to their correct storage locations.
- Throw out used gloves.
- Wash your hands with soap and water.

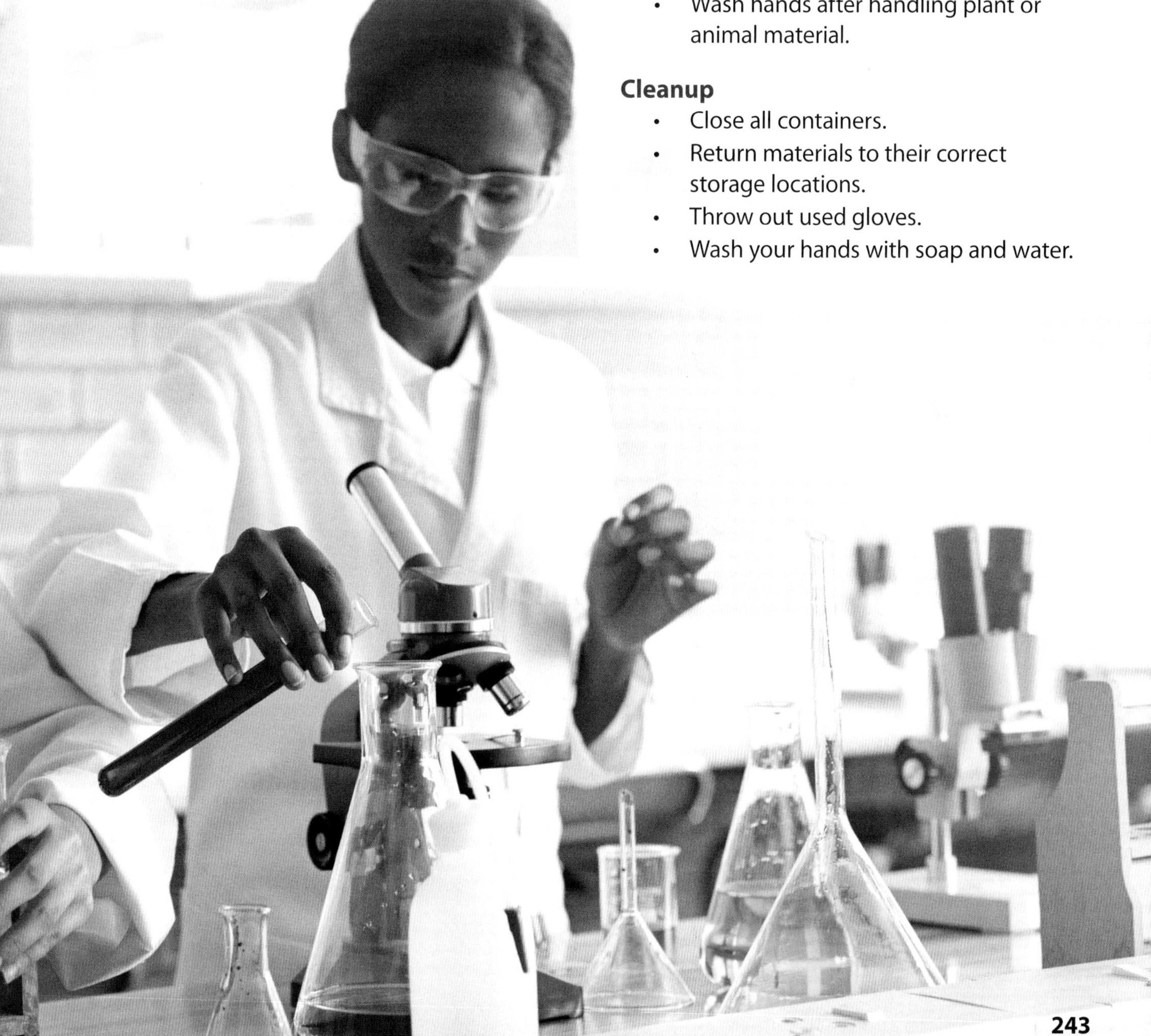

Tables and Graphs

Part of doing science involves keeping organized records of the data you collect as you investigate. Keeping detailed and careful records allows others to learn from your work or compare it to their own. Also, you can more easily make discoveries from data that are organized. A set of numbers that first appears random may reveal a meaningful pattern once it is organized. Tables and graphs are great tools for organizing and interpreting data.

Suppose you are concerned about the problem of waste disposal. You've learned that not many people recycle paper in your school. You see an empty recycling bin in the back of your art room and ask, "Would more people recycle if the recycling bin were placed closer to the garbage can?" You move the recycling bin from the back of the room and place it near the garbage can.

How will you know if moving the bin causes an increase in recycling behavior among your classmates? You need data. You plan an experiment to get the data you need. You make a plan to count the number of paper products in the recycling bin and in the garbage can at the end of each class for one week. You count the pieces of paper recycled when the recycled bin is in both positions. You jot down the numbers in your notebook. But when you try to make sense of the data, you don't remember which numbers represent which container and on which day.

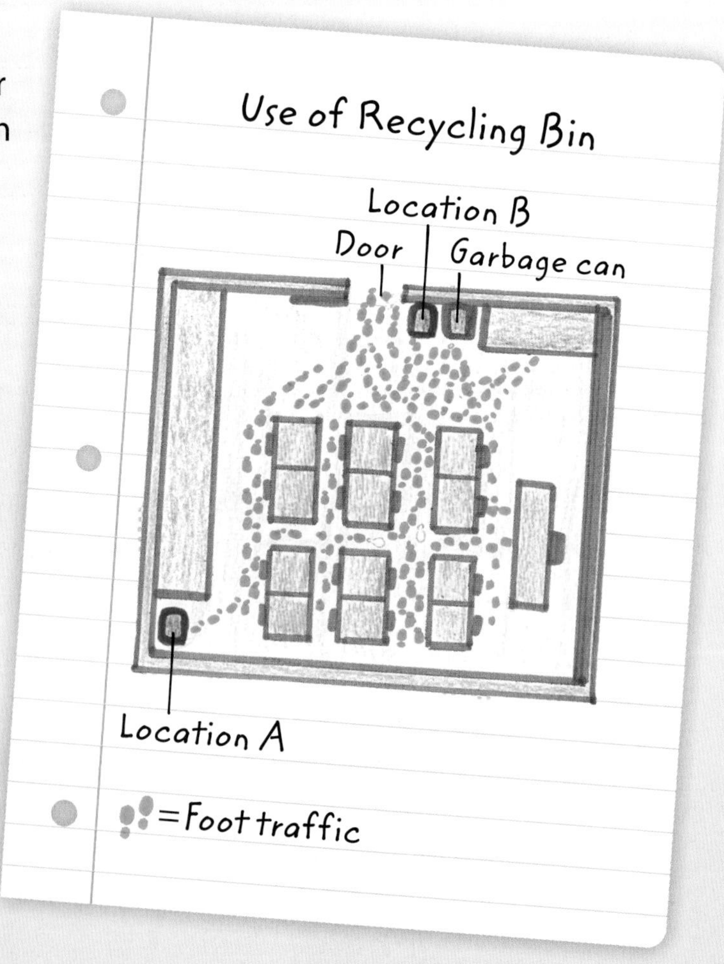

You start again. This time, you make two tables in which to carefully collect your data. A table is a set of **rows** and **columns.** Rows and columns set up a simple grid that makes a place for every piece of information.

TABLE 1. USE OF RECYCLING BIN IN BACK OF ART ROOM		
Date	**Paper Products in Recycling Bin in Location A**	**Paper Products in Garbage Can**
April 1	0	4
April 2	2	19
April 3	6	35
April 4	0	22
April 5	1	17
Total	**9**	**97**

Title

TABLE 2. USE OF RECYCLING BIN NEXT TO GARBAGE CAN		
Date	**Paper Products in Recycling Bin in Location B**	**Paper Products in Garbage Can**
April 8	15	2
April 9	41	12
April 10	9	0
April 11	18	2
April 12	5	1
Total	**88**	**17**

Column

Row

Now you have organized your data in a way that makes them useful. You can interpret the data and make claims. You can form an argument using your evidence. And you can ask more questions and make more predictions. Data are powerful when they are well organized.

Suppose you convert your totals into percentages. You can now say that only about 8 percent of the paper was recycled in the art room before you moved the bin. After the bin was moved, about 84 percent of the paper was recycled. You can make graphs to show how different these numbers are. You can argue that more recycling bins should be placed near other garbage cans in the school. And now you have the evidence to back up your argument.

Circle Graphs

Graphs provide a visual representation of information. They make it easier to see differences and similarities among data. Different types of graphs lend themselves to different types of data. A circle graph, or pie graph, is useful when you want to show the parts of a whole. For example, circle graphs are good for showing data in the form of percentages. The example circle graph identifies the six biggest sources of water consumption inside a typical home in the United States. The differences among the numbers become obvious in this format. How would the circle graph change if U.S. households eliminated all leaks?

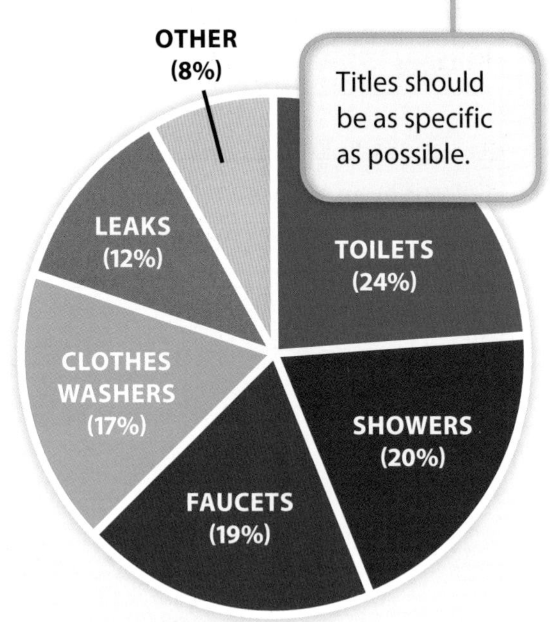

Source: Water Research Foundation, 2016

Cite the source of your data, if you did not collect if yourself. This shows others how to confirm whether your data are valid, or reliable.

Bar Graphs

Bar graphs are useful for organizing discrete data. Discrete data are data that can be counted, such as objects or events. Bar graphs are good for making comparisons among multiple similar data. For example, a student counted the number of clouds of each type observed in one month. The bar graph below shows his data. The student used colors to represent the different cloud types. Which type of cloud was observed most often? What weather changes may have affected the number of cloud types observed?

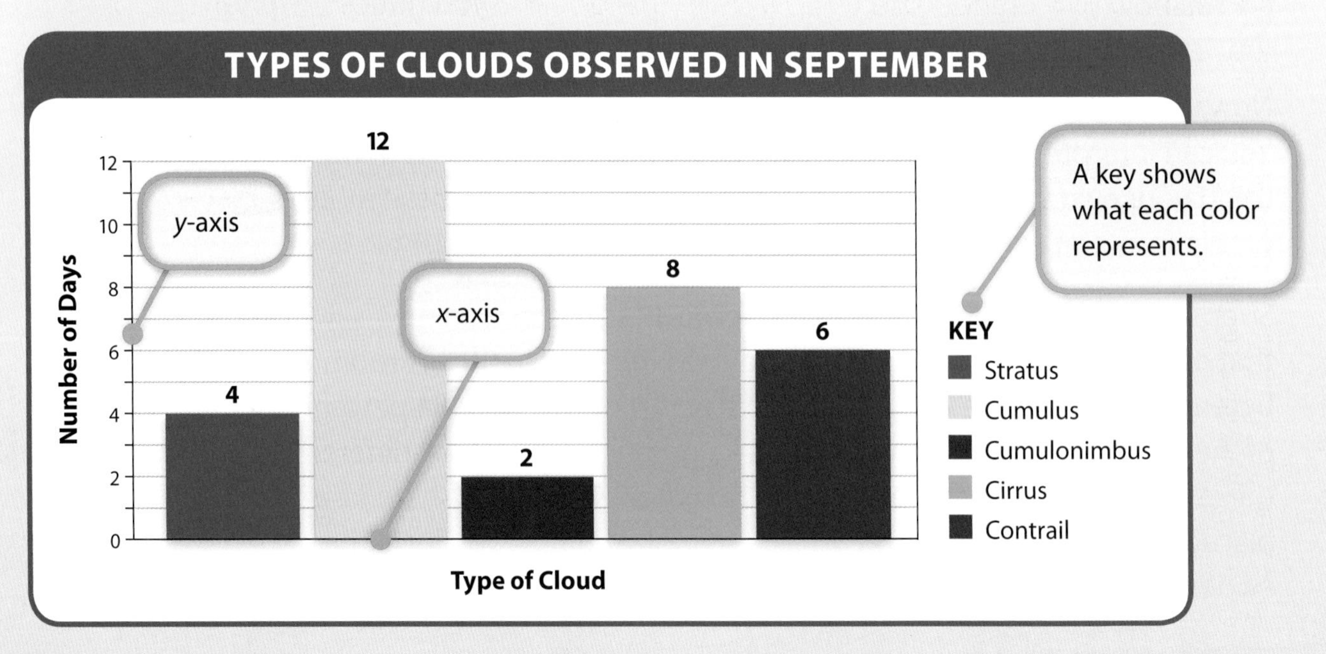

Line Graphs

Line graphs are useful for showing how data change over time. A line graph displays continuous data, or data that represent a continuous range. For example, time is continuous. You can take measurements in different units of time, such as seconds or minutes, and see a pattern of change. In a line graph, each point represents the value for a measurement. A line connects each point. Other examples of continuous data include age, distance, and speed. This line graph shows how wind speed changed over a 15-hour period in Palm Springs, California. When were winds strongest?

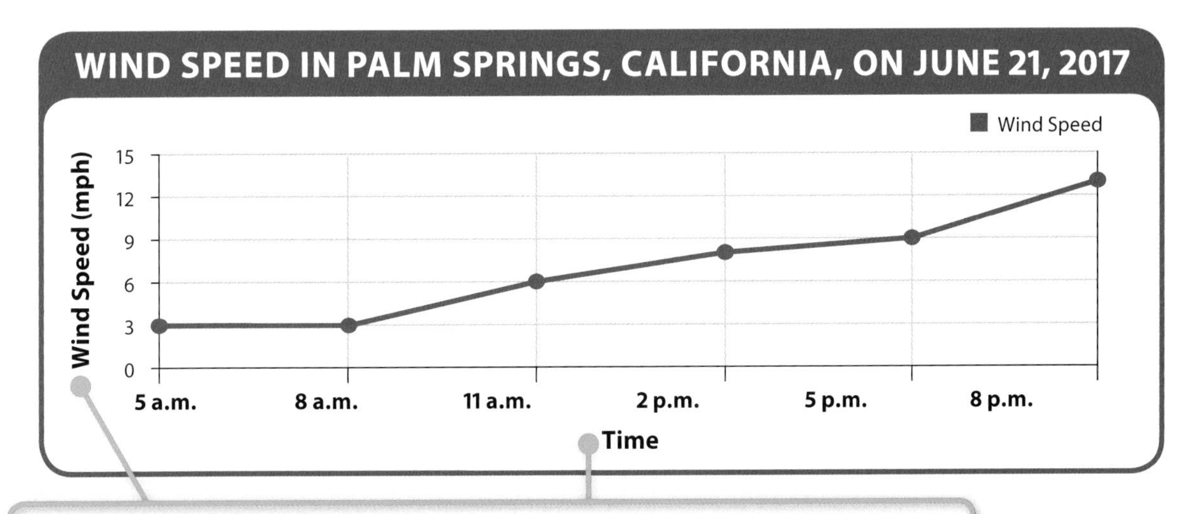

When possible, the *x*-axis and *y*-axis labels should give units of measurement.

You can add a second set of data to make a double line graph. In a double line graph, you can compare how two different variables change over the same period of time. The graph below shows how wind speed and temperature changed. Do you notice any correlation between wind speed and temperature?

A second *y*-axis is included on this side of the graph.

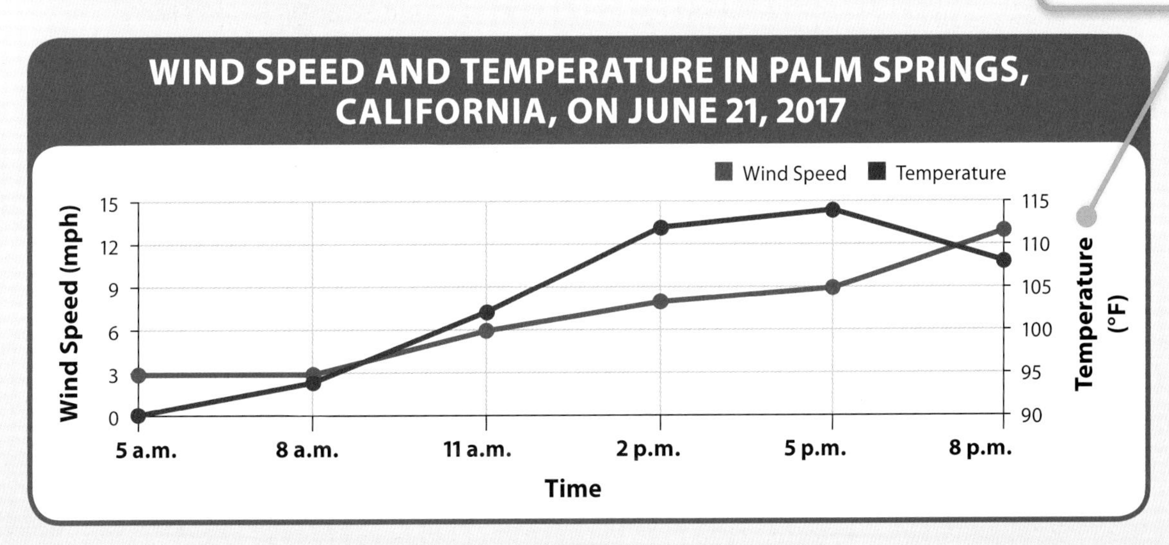

Glossary

A

agriculture (AG-ri-kul-chur)
Agriculture is the science or the activities of cultivating the soil, producing crops, and raising livestock. (p. 168)

apparent motion (uh-PAIR-ent mō-shun)
An object's apparent motion is the way it appears to move, not whether or how it actually moves. (p. 218)

atmosphere (AT-mi-sfear)
The atmosphere is the layer of air surrounding Earth. (p. 137)

axis (ak-sis)
An axis is an imaginary line that runs through the North and South poles around which the Earth spins. (p. 216)

B

bacteria (bak-TEAR-ē-uh)
Bacteria are certain kinds of one-celled living things. Bacteria are decomposers. (p. 102)

biosphere (BĪ-ō-sfear)
The biosphere is the part of Earth where life can exist. (p. 137)

boiling point (BOY-ling pôint)
At its boiling point, water begins to change from a liquid to a gas. The boiling point of water is 100°C (212°F). (p. 48)

C

chemical change (KE-mi-kul CHĀNJ)
A chemical change is a change in the state of matter in which the material becomes a different material with different properties than the original. (p. 62)

chemical reaction (KE-mi-kul rē-AK-shun)
A chemical reaction is the process by which a chemical change occurs. (p. 62)

chlorophyll (KLOR-uh-fil)
Chlorophyll is the green pigment in plants that makes it possible for them to make food from carbon dioxide and water. (p. 80)

climate (KLĪ-mit)
Climate is the pattern of the weather in an area over a long period of time. (p. 154)

column (KO-lum)
A column is a vertical section of a table. (p. 244)

community (kuh-MYŪ-nuh-tē)
A community is made up of all the different populations that live and interact in an area. (p. 109)

condensation (kon-den-SĀ-shun)
Condensation is the change from a gas to a liquid. (p. 50)

condense (kun-DENS)
To condense means to change from a gas to a liquid. (p. 50)

conservation (kon-sur-VĀ-shun)
Conservation is the protection and care of natural resources. (p. 180)

conservation of matter (kon-sur-VĀ-shun uv MA-tur)
Conservation of matter is a principle that states that the amount of matter does not increase or decrease after a reaction or change of state. (p. 49)

constellation (kon-sti-LĀ-shun)
A constellation is a group of stars that forms a particular shape in the sky and has been given a name. (p. 226)

consumer (kun-SŪ-mur)
A consumer is a living thing that eats plants or animals. (p. 94)

coral reef (KOR-ul RĒF)
A coral reef is an ocean ridge made up of coral skeletons and living coral. (p. 151)

Agriculture involves growing crops such as the ones on this terraced farm land.

The **coral reef** is a shallow ocean **ecosystem.**

D

data (DAT-a)
Data are observations and information that are recorded. (p. 10)

decomposer (dē-kum-PŌZ-ur)
A decomposer is an organism that breaks down dead organisms and the waste of living things. (p. 102)

deforestation (dē-for-is-TĀ-shun)
Deforestation is the cutting down or burning of all the trees in an area. (p. 170)

delta (DEL-tuh)
A delta is new land that forms at the mouth of a river. (p. 126)

deposition (de-pō-ZI-shun)
Deposition is the laying down of sediment and rock in a new place. (p. 153)

dissolve (di-ZAHLV)
To dissolve means to move particles of a solid throughout a liquid to form a solution. (p. 26)

E

ecosystem (Ē-kō-sis-tum)
An ecosystem is all the living things and nonliving things in an area and their interactions. (p. 107)

electrical conductor (i-LEK-tri-kul kon-DUK-ter)
An electrical conductor is a material through which electric energy can flow easily. (p. 40)

electrical conductivity
(i-LEK-tri-kul kon-duk-TIV-i-tē)
Electrical conductivity is a measure of how well electricity can move through a material. (p. 40)

electrical insulator (i-LEK-trik-ul IN-su-lā-tur)
An electrical insulator is a material that slows or stops the flow of electricity. (p. 41)

erosion (i-RŌ-zhun)
Erosion is the movement of rocks or soil caused by wind, water, or ice. (p. 153)

evaporate (i-VAP-uh-rāt)
To evaporate means to change from a liquid to a gas. (pp. 26, 140)

evidence (EV-i-dens)
A piece of evidence is an observation that supports an idea or conclusion. (p. 10)

experiment (eks-PAIR-i-ment)
In an experiment, you change only one variable, measure or observe another variable, and control other variables so they stay the same. (p. 12)

F

food web (FŪD web)
A food web is a network of food chains that shows how energy moves through an ecosystem. (p. 100)

fungi (FUN-gī)
Fungi are organisms such as molds and mushrooms, that have no chlorophyll and live on dead and decaying things. Fungi are decomposers. (p. 102)

G

gas (GAS)
Gas is matter that spreads to fill a space. (p. 24)

geosphere (JĒ-ō-sfear)
The geosphere is the solid outer part of Earth composed of rock and thought to be about 60 miles thick. (p. 136)

glacier (GLĀ-shur)
A glacier is a huge, slow-moving mass of ice. (p. 162)

gravity (GRA-vi-tē)
Gravity is a force that pulls objects toward each other. (p. 198) *also* **gravitational force** (GRA-vi-TĀ-shun-ul FORS) (p. 203)

groundwater (GROWND-wah-tur)
Groundwater is water held underground in spaces within soil and rock. (p. 140)

H

hardness (HARD-nis)
Hardness is a measure of how resistant a material is to scratching, bending, or denting. (p. 34)

hydroponics (hī-drō-PAH-niks)
Hydroponics is a method of growing plants in water instead of soil. (p. 86)

hydrosphere (HĪ-drō-sfear)
The hydrosphere is all the water at or near Earth's surface, including liquid bodies of water, frozen water as ice and snow, water found underground, and water vapor in the atmosphere. (p. 136)

hypothesis (hī-POTH-uh-sis)
A hypothesis is a statement giving a possible answer to a question that can be tested by an experiment. (p.12)

I

Infer (in-FUR)
When you infer, you use what you know and what you observe to draw a conclusion. (p. 10)

invasive species (in-VĀ-siv SPĒ-shēz)
An invasive species is a species that has been brought to a new place by people and can harm the environment. (p. 119)

investigate (in-VES-ti-gāt)
You investigate when you carry out a plan to answer a question. (p. 12)

L

liquid (LIK-wid)
A liquid is matter that takes the shape of its container. (p. 24)

M

magnetism (MAG-nuh-ti-zum)
Magnetism is a force between magnets and objects magnets attract. (p. 39)

mass (MAS)
Mass is the amount of matter in an object. (p. 22)

matter (MA-tur)
Matter is anything that takes up space. (p. 22)

melting point (MELT-ing POYNT)
At its melting point, frozen water will change into a liquid. The melting point of water is warmer than 0°C (32° F). (p. 48)

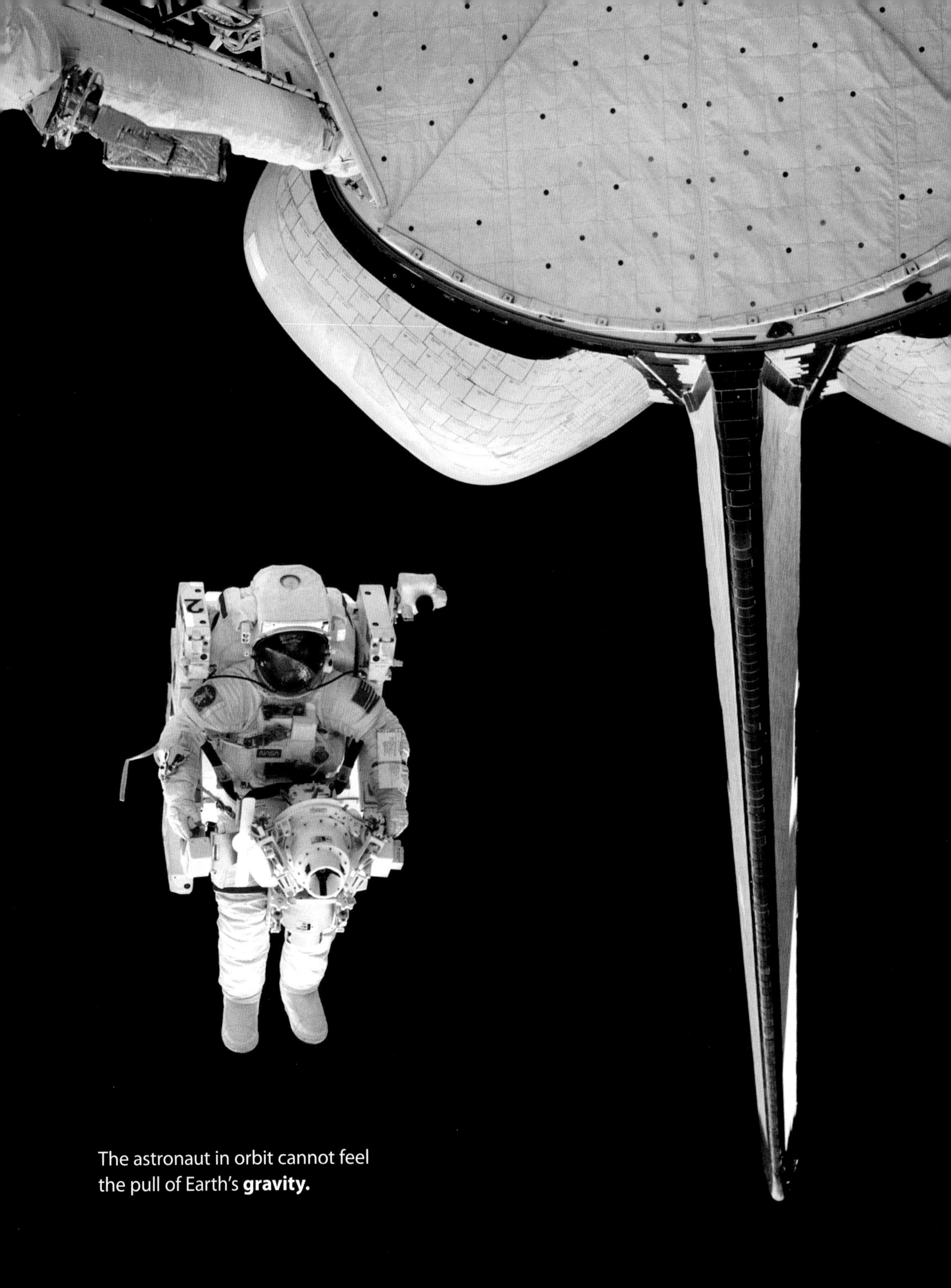

The astronaut in orbit cannot feel the pull of Earth's **gravity.**

mixture (MIKS-chur)
A mixture is two or more kinds of matter put together. (p. 58)

model (MO-del)
In science, models are used to explain or predict phenomena. A model can show how a process works in real life. (p. 12)

monsoon (mon-SŪN)
A monsoon is a seasonal weather pattern of a wet and dry season. (p. 146)

N

nonrenewable energy resource
(non-rē-NŪ-i-bul EN-ur-jē RĒ-sors)
A nonrenewable energy resource is an energy resource that cannot be replaced quickly enough to keep from running out. (p. 188)

nonrenewable resource
(non-rē-NŪ-i-bul RĒ-sors)
A nonrenewable resource is a resource that cannot be replaced quickly enough to keep from running out. (p. 167)

nutrient (NŪ-trē-ent)
A nutrient is a part of food and soil that helps living things stay healthy and grow. (p. 82)

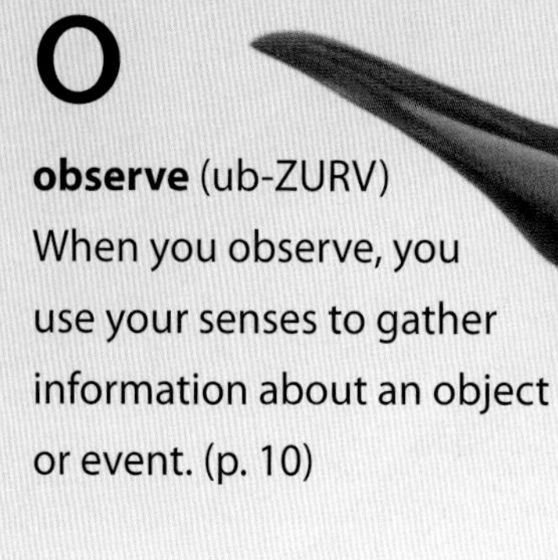

O

observe (ub-ZURV)
When you observe, you use your senses to gather information about an object or event. (p. 10)

ocean current
(Ō-shun KUR-ent)
An ocean current is a steady flow of ocean water. (p. 152)

orbit (OR-bit)
An orbit is a path a revolving body follows. (p. 222)

A group of individual king penguins living in the same area forms the **population**.

P

phase (FĀZ)
The moon's phase is how the moon appears from Earth. (p. 232)

photosynthesis (fō-tō-SIN-thi-sis)
Photosynthesis is the chemical process that green plants use to turn water and carbon dioxide into food when the plant is exposed to light. (p. 80)

physical change (FI-zi-kul CHĀNJ)
A physical change is a change in the state of matter in which the material is still the same type of material with the same properties as the original. (p. 48)

population (pop-yū-LĀ-shun)
A population is all the individuals of a species that live in an area. (p. 108)

producer (pruh-DŪS-ur)
A producer is a living thing that makes its own food. (p. 94)

property (PRAH-pur-tē)
A property is something about an object that you can observe with your senses. (p. 32)

R

rain shadow (RĀN SHA-dō)
A rain shadow is a region of reduced rainfall on the side of high mountains that is sheltered from the wind. (p. 157)

recycle (rē-SĪ-kul)
To recycle is to use materials in an old object to make a new object. (p. 181)

renewable energy resource
(rē-NŪ-i-bul EN-er-jē RĒ-sors)
A renewable energy resource is an energy resource that is continually replaced and will not run out. (p. 189)

renewable resource (rē-NŪ-i-bul RĒ-sors)
A renewable resource is a resource that is always being replaced and will not run out. (p. 166)

revolve (rē-VAHLV)
To revolve is to travel around another object in space. (p. 202)

revolution (re-vuh-LŪ-shun)
A revolution is the motion of one object around another object. (p. 203)

rotate (RŌ-tāt)
To rotate is to spin around. (p. 216)

row (RŌ)
A row is a horizontal section of a table. (p. 244)

runoff (RUN-awf)
Runoff are the materials, such as nutrients, picked up and carried by flowing rainwater. (p. 174)

S

science communication
(SĪ-ens kuh-MYŪ-nuh-KĀ-shun)
Science communication is the sharing of science ideas with others. (p. 238)

smog (SMAHG)
Smog is a thick haze caused by the action of sunlight on air polluted by smoke or automobile exhaust fumes. (p. 176)

solid (SAHL-id)
A solid is matter that has its own shape. (p. 24)

solution (suh-LŪ-shun)
A solution is a mixture of matter dissolved in a liquid. (p. 46)

solubility (SOL-yū-BIL-uh-tē)
Solubility is the ability of matter to dissolve in a liquid. (p. 46)

space junk (SPĀS JUNK)
Space junk is material made by humans that has been left in orbit around Earth. (p. 178)

species (SPĒ-shēz)
A species is a group of similar living things that can produce offspring who can, in turn, produce offspring. (p. 108)

star (STAR)
A star is a ball of hot gases that gives off light and other types of energy. (p. 212)

states of matter (STĀTS uv MA-tur)
States of matter are the forms in which a material can exist. (p. 24)

T

thermal conductor (THUR-mul kun-DUK-ter)
A thermal conductor is a material that heats up quickly. (p. 44)

thermal energy (THUR-mul EN-ur-jē)
Thermal energy is the energy of heat. (p. 44)

thermal insulator (THUR-mul IN-suh-lā-tur)
A thermal insulator is a material that heats up slowly. (p. 45)

V

variable (VAIR-ē-u-bl)
A variable is a factor that can change or be controlled in an experiment, investigation, or model. (p. 12)

vegetation (vej-uh-TĀ-shun)
Vegetation is the plant life or all the plants that cover a particular area. (p. 170)

vertical forest (VUR-ti-kul FOR-ist)
A vertical forest is a tall building on which many trees and other plants are grown. (p. 186)

W

weather (WE-thur)
Weather is the state of the atmosphere at a certain place and time. (p. 154)

X

***x*-axis** (EKS AKSIS)
The *x*-axis on a graph is the horizontal base line. The *x*-axis is usually presented on the bottom of the graph. (p. 246)

Y

***y*-axis** (WI AKSIS)
The *y*-axis on a graph is the vertical base line. The *y*-axis is usually presented on the left side of the graph. (p. 246)

Lightning is one sign of severe **weather.**

Index

A

B

C

D

E

I

J

K

L

M

T

U

V

W

Z

Photographic and Illustrator Credits

Front Matter

Title page ©Mimi Ditchie Photography/Getty Images. **ii-iii** (c) ©Aaron Joel Santos/Getty Images. **iv-v** (c) ©Joel Sartore/Getty Images. **vi-vii** (c) ©arlindo71/Getty Images. **viii-ix** (c) ©shannonstent/Getty Images. **x-xi** (c) ©agoncalves/Getty Images.

Welcome to Exploring Science

2 (tl) ©Sofía Ruzo. **2-3** ©Devlin Gandy/National Geographic Creative. **4** (bl) ©Andrés Ruzo. (bc) ©David Salazar Photography. **7** ©Devlin Gandy/ National Geographic Creative.

Nature of Science

8-9 (c) ©Science History Images/Alamy Stock Photo. **10** (bc) ©Turtle Rock Scientific/Science Source. **10-11** (c) ©Elizabeth Greer/EyeEm/ Getty Images. **12** (bc) ©John van Hasselt - Corbis/Getty Images. **12-13** (c) ©NASA/Science Source. **14** (b) ©Brian Dexter/Getty Images. **15** (bc) ©Jeff Fitlow/Rice Uinversity. **16** (tl)(tr)(bl)(bc) ©Courtney Rader/ National Geographic Learning/Cengage. (tc) ©National Geographic School Publishing. **16-17** (c) ©SPL/Science Source. **17** (tl) ©Courtney Rader/National Geographic Learning/Cengage. (br) ©Courtney Rader/ National Geographic Learning/Cengage. **18** (tr) ©Sofía Ruzo. **18-19** ©Rico Ködder/EyeEm/Getty Images.

Physical Science: Structure and Properties of Matter

20-21 (c) ©Andrew Winning/Reuters. **22-23** (c) ©Philip Lange/ ShutterStock.com. **23** (cl) ©Gregory G. Dimijian, M.D./Science Source. (cr) ©Raul Gonzalez Perez/Science Source. **24** (bl) ©Precision Graphics/National Geographic Learning. (br) ©Precision Graphics/ National Geographic Learning. **24-25** (c) ©Design Pics/Design Pics/ Superstock. **25** (tl) ©Precision Graphics/National Geographic Learning. (tr) ©Michael Goss Photography/National Geographic Learning. (cr) ©Michael Goss Photography/National Geographic Learning. **26** (tl) (tc)(tr)(bl)(br) ©National Geographic Learning. (bc) ©Michael Goss Photography/National Geographic Learning. **26-27** (c) ©Peter Dazeley/ Photographer's Choice/Getty Images. **27** (tl) ©Jennifer Shaffer/National Geographic Learning. (tr) ©Michael Goss Photography/National Geographic Learning. **28-29** (c) ©Dan Herrick/Lonely Planet Images/ Getty Images. **30** (tl) ©Joseph Sohm/Getty Images. (tr) ©asharkyu/ Shutterstock.com. (bc) ©EPA/MICHAEL REYNOLDS/Redux Pictures. **31** (tr) Janette Pellegrini/Getty Images. **32** (c) ©Chris Scredon/E+/Getty Images. **33** (tl) ©Chris Pancewicz/Alamy Stock Photo. (tr) ©Photodisc/ Getty Images. (cr) ©Ginette Chapman/Arcaid/Corbis. (cl) ©DAJ/Getty Images. (bl) ©Chad Zuber /Shutterstock.com. **34** (cl) ©Ozgur Coskun/ Shutterstock.com. **34-35** (c) ©Peter de Clercq/Alamy Stock Photo. **35** (c) ©National Geographic Learning. **35** (t)(cl)(c)(bl)(br) ©National Geographic Learning. (cr) ©Jennifer Shaffer/National Geographic School Publishing. **36** (t) ©Michael Goss Photography/National Geographic Learning. (bl) ©Jennifer Shaffer/National Geographic School Publishing. (bc) ©National Geographic School Publishing. (bl) ©National Geographic Learning. **36-37** (c) ©Melissa Carroll/E+/ Getty Images. **37** (tl)(tr)(bl)(br) ©Michael Goss Photography/National Geographic Learning. **38-39** (c) ©Bernd Mellmann/F1online/Getty Images. **39** (tr) ©National Geographic Learning. **40-41** (c) ©Igor Ostapchuk/Shutterstock.com. **41** (tl) ©Tunart/iStockphoto.com. (tr) ©Yury Kosourov /Shutterstock.com. (bl) ©Kelah2001/iStockphoto.com. (br) ©ID1974 /Shutterstock.com. **42** (tl)(tc)(tl)(bl)(bc)(br) ©National Geographic Learning. **42-43** (c) ©klenger/E+/Getty Images. **43** (tl1)(tl2)(tl3)(tl4) ©National Geographic Learning. **43** (tr)(cl) ©Michael Goss Photography/National Geographic Learning. **44** (cl) ©Stefano Scata/Photolibrary/Getty Images. (c) ©chinamols/iStockphoto.com. (cr) ©Dave King/Dorling Kindersley/ Getty Images. **44-45** (c) ©Sami Sarkis/Photographer's Choice RF/ Getty Images. **45** (cl) ©Thomas Collins/Photographer's Choice/Getty Images. (tr)(cr) ©Michael Goss Photography/National Geographic Learning. **46** (tr)(tl)(bc2)(br) ©Jennifer Shaffer/National Geographic Learning. (bc1)(bl) ©National Geographic Learning. **46-47** (c) ©Tohru minowa/a.collectionRF/Getty Images. **47** (tl)(tr) ©Jennifer Shaffer/ National Geographic School Publishing. **48-49** (c) ©Martin Dohrn/ Science Source. **49** (c) ©Steve Holroyd/Alamy Stock Photo. **50-51** (c) ©LianeM/Alamy Stock Photo. **51** (tr) ©istockphoto.com/Razvan . (cr) ©Adam Clark/Aurora/Getty Images. **52-53** ©Henn Photography/ Getty Images. **54** (tc) ©Greg Dale/National Geographic Creative. **55** ©Cordelia Molloy/Science Source. **56** (tl)(tc)(tr)(bl)(bc)(br) ©National Geographic School Publishing. **56-57** (c) ©Yanika Panfilova/Alamy Stock Photo. **57** (tl) ©National Geographic School Publishing. (tr) ©Michael Goss Photography/National Geographic Learning. **58** (tl) (tc)(tr)(bl)(bc)(br) ©National Geographic Creative. **58-59** (c) ©HaiLight Fotografie/StockFood/Age Fotostock. **59** (tl)(tr) ©Michael Goss Photography/National Geographic Learning. **60-61** (c) ©josemoraes/ E+/Getty Images. **62-63** (c) ©Lacaosa/Moment/Getty Images. **63** (t) (c)(b) ©Precision Graphics. **64-65** (c) ©Jeffrey Coolidge/Iconica/Getty Images. **65** (tl) ©Darren Robb/The Image Bank/Getty Images. (tr) ©Charles D. Winters/Science Source. **66** (c) ©National Geographic Learning. **66** (tr) ©National Geographic Learning. **66** (tl)(tc)(bl)(bc)(br) ©National Geographic School Publishing. **66-67** (c) ©Colin Anderson/ Stockbyte/Getty Images. **67** (tl)(tr) ©Michael Goss Photography/ National Geographic Learning. **68-69** (c) ©R. creations/amanaimages/ Corbis. **70** (br) ©Erik Jepsen. **70-71** (c) ©Erik Jepsen. **71** (tr) ©Courtesy of Albert Lin. **73** (bc) ©Devlin Gandy. **74** (tr) ©Sofía Ruzo. **74-75** (spread) ©Zoonar/Elena Elissee/AGE Fotostock.

Life Science: Matter and Energy in Organisms and Ecosystems

76-77 (c) ©Martin Harvey/Photolibrary/Getty Images. **78** (tr) ©John Anderson/Alamy Stock Photo. **78-79** (c) ©MG photo studio/Alamy Stock Photo. **80-81** (c) ©Sergio Del Rosso Photography/Moment Open/Getty Images. **81** (t) ©Precision Graphics. **82-83** (c) ©skywing/ Shutterstock.com. **83** (cr) ©Kimiko Ichikawa/Flickr/Getty Images. **84** (tr) ©Dervin Witmer/Shutterstock.com. **84-85** (c) ©Raul Touzon/ National Geographic Creative. **85** (tl) ©Joel Sartore/National Geographic Creative. (tr) ©Peter Essick/National Geographic Creative. **86** (br) ©Galen Rowell/Mountain Light/Alamy Stock Photo. (bl) ©Felipe Pederos-Bustos/National Science Foundation. **86-87** (c) ©Paul S. Howell/Liaison/Hulton Archive/Getty Images. **88** (bl)(bc1) (br) ©Jeanine Childs/National Geographic Learning. (bc2) ©National Geographic Learning. **88-89** (c) ©Chanyut Sribua-rawd/E+/Getty Images. **89** (tl)(tr) ©Jeanine Childs/National Geographic Learning. **90-91** (c) ©Bplanet/Shutterstock.com. **92** (tr) ©Oliver Strewe/Lonely Planet Images/Getty Images. **92-93** (c) ©Frans Lanting/National Geographic Creative. **94** (cl) ©BlooMimage/Getty Images. (c) ©Brendan Bucy/ Shutterstock.com. (cr) ©Barry Mansell/Nature Picture Library. **94-95** (c) ©Ed Darack/Science Faction/Getty Images. **95** (cl) ©Rusty Dodson/ Shutterstock.com. (cr) ©James McLaughlin/Alamy Stock Photo. **96** (tr) © Arthur Tilley/Creatas/JupiterImages. (cr) ©Arthur Tilley/Creatas/ JupiterImages. **96-97** (t) ©John Foxx/Stockbyte/Getty Images. (b)

©Digital Vision/Getty Images. **97** (tl) ©Wim van Egmond/Visuals Unlimited/Getty Images. (tc) ©Scenics and Science/Oxford Scientific/Getty Images. (tr) ©George Grall/National Geographic Creative. (bl) ©Ammit/Alamy Stock Photo. (bc) ©Christian Musat/Alamy Stock Photo. (br) ©Steve Winter/National Geographic Creative. **98-99** (c) ©Richard Du Toit/Minden Pictures. **100** (cl) ©DLILLC/Cardinal/Corbis. (tr) ©Jeff Vanuga/Flame/Corbis. (br) ©Chris Howes/Wild Places Photography/Alamy Stock Photo. **100-101** (c) ©Tim Fitzharris/Minden Pictures. **101** (tl) ©Geostock/Photodisc/Getty Images. (c) ©Jack Goldfarb/Design Pics/Corbis. (bl) ©Rich Reid/National Geographic Creative. (cr) ©Rich Reid/National Geographic Creative. **102-103** (c) ©TJ Blackwell/Flickr/Getty Images. **103** (tr) ©Eye of Science/Science Source. **104-105** (c) ©prettyfoto/Alamy Stock Photo. **106-107** (c) ©Wolfgang Kaehler/SuperStock. **107** (tr)(cr) ©Jim Brandenburg/Feature Stories/American Prairie/Minden Pictures. **108** (cl) ©Dick Kettlewell/America 24-7/Getty Images. (cr) ©Raymond Gehman/National Geographic Creative. **108-109** (c) ©Jim Brandenburg/Feature Stories/American Prairie/Minden Pictures. **109** (cl) ©Raymond Gehman/National Geographic Creative. (cr) ©Jim Brandenburg/Feature Stories/American Prairie/Bison/Minden Pictures. **110** (tl)(tc1)(tc2)(tl)(bl)(bc1)(bc2)(br) ©National Geographic School Publishing. **110-111** (c) ©Jerome Wexler/Science Source. **111** (tl)(tr)(cl) ©National Geographic School Publishing. **112-113** (c) ©Mitsuhiko Imamori/Minden Pictures. **114-115** ©FLPA/Alamy Stock Photo. **116** ©ZUMA Press, Inc./Alamy Stock Photo**. 117** ©Handout/MCT/Newscom. **118-119** (c) ©Rob Hainer/Shutterstock.com. **119** (cr) ©Mapping Specialists. **120-121** (c) ©Scott Leslie/Minden Pictures. **121** (tr) ©Mapping Specialists. **122-123** (c) ©John Abbott/Visuals Unlimited/Getty Images. **123** (cr) ©Sanford Porter/Agricultural Research Service, USDA. (tl) ©Precision Graphics. (tr) ©Sanford Porter/Agricultural Research Service, USDA. **124** (tl)(tr)(bc) ©Tree Foundation. **126-127** (c) ©Peter McBride/National Geographic Creative. **127** (bl) ©Mapping Specialists. (tr) ©Rebecca Hale/National Geographic Creative. **128-129** (c) ©Agencia el Universal/El Universal de Mexico/Newscom. **129** (t) ©Agencia el Universal/El Universal de Mexico/Newscom. **131** ©Sofía Ruzo**. 132-133** ©Nuno Filipe Pereira/EyeEm/Getty Images.

Earth Science: Earth's Systems/Space Systems: Stars and the Solar System

134-135 (c) ©Jason Corneveaux/Flickr Open/Getty Images. **136** (t) ©Rolf Hicker/All Canada Photos/Getty Images. (b) ©John Eastcott and Yva Momatiuk/National Geographic Creative. **136-137** (c) ©NASA Goddard Space Flight Center. **137** (t) ©NASA. (b) ©Frans Lanting/National Geographic Creative. **138** (tr1) ©NASA Goddard Space Flight Center. (tr2) ©Rolf Hicker/All Canada Photos/Getty Images. (cl) ©James L. Amos/National Geographic Creative. (tr3) ©John Eastcott and Yva Momatiuk/National Geographic Creative. (tr4) ©NASA. (tr5) ©Frans Lanting/National Geographic Creative. **138-139** (c) ©Rolf Hicker/All Canada Photos/Getty Images. **139** (tr) ©George Clerk/E+/Getty Images. **140** (tr1) ©NASA Goddard Space Flight Center. (tr2) ©Rolf Hicker/All Canada Photos/Getty Images. (tr3) ©John Eastcott and Yva Momatiuk/National Geographic Creative. (tr4) ©NASA. (tr) ©Frans Lanting/National Geographic Creative. **140-141** (c) ©John Eastcott and Yva Momatiuk/National Geographic Creative. **142** (tr1) ©NASA Goddard Space Flight Center. (tc) ©Rolf Hicker/All Canada Photos/Getty Images. (tr) ©John Eastcott and Yva Momatiuk/National Geographic Creative. (tr) ©NASA. (cl) ©Precision Graphics. (tr) ©Frans Lanting/National Geographic Creative. **142-143** (c) ©NASA. **144** (tr) ©NASA Goddard Space Flight Center. (tr) ©Rolf Hicker/All Canada Photos/Getty Images. (tr) ©John Eastcott and Yva Momatiuk/National Geographic Creative. (tr) ©NASA. (tr) ©Frans Lanting/National Geographic Creative. **144-145** (c) ©Frans Lanting/National Geographic Creative. **145** (c) ©Paul Chesley/National Geographic Creative. **146** (tr) ©NASA Goddard Space Flight Center. (tr) ©Rolf Hicker/All Canada Photos/Getty Images. (tr) ©John Eastcott and Yva Momatiuk/National Geographic Creative. (tr) ©NASA. (tr) ©Frans Lanting/National Geographic Creative. **146-147** (c) ©Parvez Khaled Photography/Flickr/Getty Images. **147** (tr) ©James P. Blair/National Geographic Creative. (cr) ©Mustafiz Mamun/Majority World/Age Fotostock. **148** (bc1)(tl)(tc)©Michael Goss Photography/National Geographic Learning.(bc2)(bl)©National Geographic School Publishing. (br)©National Geographic Learning. (tr)©Jeanine Childs/National Geographic Learning. **148-149** (c) ©catherine raillard/Corbis. **149** (tl)(tr)(cl) ©Michael Goss Photography/National Geographic Learning. **150-151** (c) ©Alexander Safonov/Flickr Select/Getty Images. **151** (t) ©David Tipling/Getty Images. (c) ©Georgette Douwma/Photographer's Choice/Getty Images. (b) ©David Shale/Nature Picture Library. **152** (tl) ©Spring Images/Alamy Stock Photo. (tr) ©Mapping Specialists. **152-153** (c) ©Frans Lanting/National Geographic Creative. **154** (tr) ©Tony Arruza/Corbis. **155** (c) ©Raven/Science Source. **156-157** (c) ©NASA. **158** (tr) ©Mapping Specialists. **158-159** (c) ©Florian Sprenger/Flickr/Getty Images. **160-161** (c) ©Georgette Douwma/The Image Bank/Getty Images. **162-163** (c) ©Steve Allen/The Image Bank/Getty Images. **163** (tr) ©Oliver Strewe/Lonely Planet Images/Getty Images. (cr) ©Paul Harris/AWL Images/Getty Images. **164** (cl) ©Michael Goss Photography/National Geographic Learning. **164-165** (c) ©Panoramic Images/Getty Images. (cl) ©istockphoto.com/Steve Cole. (c) ©iStockphoto/David Sucsy. (c) ©Dex Image/Alamy Stock Photo. (cr) ©David R. Frazier/Science Source. **166-167** (c) ©Peter Miller/Getty Images. (cl) ©Raymond Gehman/National Geographic Creative. (c) ©Vadim Ponomarenko/Shutterstock.com. (cr) ©Neil Beer/Photodisc/Getty Images. **168-169** (c) ©Jim Wark/Agstockusa/AGE Fotostock. **169** (cr) ©Hans-Peter Merten/The Image Bank/Getty Images. (tr) ©Superstudio/Iconica/Getty Images. **170-171** (c) ©Cardinal/Corbis. **171** (cr) ©Oliver Strewe/Stone/Getty Images. **172** (tl)(tc1)(tc2)(bc1)(bc2)(br) ©National Geographic School Publishing. (tr1)(bl) ©Michael Goss Photography/National Geographic Learning. **172-173** (c) ©Dimitri Otis/Photographer's Choice/Getty Images. **173** (cl)(cr) ©Michael Goss Photography/National Geographic Learning. **174-175** (c) ©Visage/Stockbyte/Getty Images. **175** (cr) ©Michael Patrick O'Neill/Science Source. **176** (tr) ©Mapping Specialists. **176-177** (c) ©fototrav/E+/Getty Images. **177** (c) ©Paolo Patrizi/Alamy Stock Photo. **178-179** (c) ©NASA Orbital Debris Office. **179** (tl) ©Orange County Register/Zumapress/Newscom. (tr) ©Anna Zieminski/afp/Newscom. **180** (cl) ©Izabela Habur/iStockphoto.com. (c1) ©Wave Royalty Free/Science Source. (c2) ©Bob Daemmrich/Alamy Stock Photo. (cr) ©Tim Pannell/Ivy/Corbis. **180-181** (c) ©Andia/Alamy Stock Photo. **181** (tr)(cr) ©National Geographic School Publishing. **182** (tl) ©George Rinhart/Getty Images. (tr) Everett Collection Inc/Alamy Stock Photo. (bc) Alfred Eisenstaedt/Getty Images. **183** ©Stock Montage/Getty Images. **184** (t c) ©Mapping Specialists. **184-185** (c) ©Ian Homer/Alamy Stock Photo. **186** (bc) ©Marco Garfalo. **186-187** (c) ©Iberpress/ZUMAPRESS/Newscom. **188-189** (c) ©imagebroker/Alamy Stock Photo. **189** (tr) ©bl Images Ltd/Alamy Stock Photo. (cr) ©Imaginechina/Corbis. **190** ©Andrea Altemueller/Getty Images. **190-191** ©Harald Sund/Getty Images. **192-193** ©David Nunuk/Getty Images. **194** (tl)(bc3)(bl)©Michael Goss

Photography/National Geographic Learning.(tc1)(tc3)(tl) (bc1)©National Geographic School Publishing. (tc2)(bc2)©National Geographic Learning.(bl)©Jeanine Childs/National Geographic Learning. **194-195** (c) ©Albaimages/Alamy Stock Photo. **195** (tl) ©Michael Goss Photography/National Geographic Learning. **195** (cl)(tr) ©Jeanine Childs/National Geographic Learning. **196-197** (c) ©Jeff Greenberg1of 6/Alamy Stock Photo. **198** (cl) ©Precision Graphics. **198-199** (c) ©Buzz Pictures/Alamy Stock Photo. **199** (tr) ©Oliver Furrer/ cultura/Corbis. **200** (cl) ©National Geographic School Publishing. (c) ©National Geographic Learning. (bc) ©Michael Goss Photography/ National Geographic Learning. (bl) ©Jennifer Shaffer/National Geographic School Publishing. (bc) ©National Geographic Learning. **200-201** (c) ©Borut Trdina/E+/Getty Images. **201** (tr)(bl)(br) ©Michael Goss Photography/National Geographic Learning. **202-203** (c) ©Precision Graphics. **204** (bc) ©Courtesy of NASA. (tl)(tc)(cl)(c) ©CosmoQuest.org. **204-205** ©Courtesy of NASA/GSFC/Arizona State University. **206-207** (c) ©Courtesy of NASA. **208** (tl) ©sevenke/ Shutterstock. (tc) ©Andy Piatt/Shutterstock. **208-209** (t) ©irin-k/ Shutterstock. **209** (tr) ©Brian Nehlsen/National Geographic Learning/ Cengage. (tc) ©kai keisuke/Shutterstock. **210** (tl)(tr)(bc) ©Courtesy of NASA. **211** ©Courtesy of NASA. **212-213** (c) ©jtb Photo/SuperStock. **213** (tl) ©Jim Richardson/National Geographic Creative. **213** (tr)(inset)(cr) ©Jeanine Childs/National Geographic Learning. **214** (tl)(tc)(tr)(bc) ©National Geographic Learning. **214-215** (c) ©amana images inc./ Alamy Stock Photo. **215** (tl) ©Precision Graphics. (cl)(cr) ©Michael Goss Photography/National Geographic Learning. **216** (bl)(inset) ©National Geographic School Publishing. **216** (inset) ©National Geographic School Publishing. **216-217** (c) ©shulz/iStockphoto.com. **217** (cr) ©National Geographic School Publishing. **218-219** (c) ©Herbert Hopfensperger/Age Fotostock. **219** (tr) ©Ben Senior and Shanin Glenn/ National Geographic School Publishing. **220** (tl)(tc)(tr)(bc) ©National Geographic School Publishing. **220-221** (c) ©Mark Leman/ Photographer's Choice/Getty Images. **221** (tl)(tr) ©National Geographic School Publishing. **222-223** (c) ©Lasse Kristensen/Alamy Stock Photo. **223** (tr)(cr1)(cr2)(br) ©Gary Buss/Taxi/Getty Images. (ct)(c1)(c2)(cb) ©Precision Graphics. **224** (cl) ©Michael Goss Photography/National Geographic Learning. **224-225** (c) ©jeremyvandermeer/Flickr Open/ Getty Images. **225** (tl)(tr) ©Precision Graphics. **226-227** (c) ©Precision Graphics. **227** (t) ©Precision Graphics. **228-229** (c) ©Thomas Backer/ Aurora/Getty Images. **229** (t) ©Precision Graphics. **230** (bl) ©George F. Mobley/National Geographic Creative. **230-231** (c) ©Tom Kelly Photo/ Flickr/Getty Images. **232-233** (c) ©pjmorley/Shutterstock. **233** (t) ©Precision Graphics. **234** (bl)(bc)(br)(cl) ©National Geographic School Publishing. (cr) ©Precision Graphics. **234-235** (c) ©DigitalStock/Corbis. **235** (tl)(tr) ©Jennifer Shaffer/National Geographic School Publishing. **236-237** (c) ©Rebecca Hale/National Geographic Creative. **237** (tr) ©Henry Throop. **238** (tl) ©Brendan Mullan. (tr) ©Brendan Mullan. **238-239** (c) ©Brendan Mullan. **241** ©Devlin Gandy. **242-243** (c) ©Klaus Vedfelt/Getty Images.

Glossary

248-249 (c) ©Keren Su/China Span/Getty Images. **250-251** (c) ©Borut Furlan/WaterFrame/Getty Images. **252-253** (c) ©Terra/Corbis. **254-255** (c) ©John Eastcott and Yva Momatiuk/Stone/Getty Images. **256-257** (c) ©Boris Jordan Photography/Flickr/Getty Images.

Illustration Credits

Unless otherwise indicated, all illustrations were created by Lachina, and all maps were created by Mapping Specialists.

Program Consultants

Randy L. Bell, Ph.D.
Associate Dean and Professor of Science Education, College of Education, Oregon State University

Malcolm B. Butler, Ph.D.
Professor of Science Education and Associate Director; School of Teaching, Learning and Leadership; University of Central Florida

Kathy Cabe Trundle, Ph.D.
Department Head and Professor, STEM Education, North Carolina State University

Judith S. Lederman, Ph.D.
Associate Professor and Director of Teacher Education, Illinois Institute of Technology

Center for the Advancement of Science in Space, Inc.
Melbourne, Florida

Acknowledgments
Grateful acknowledgment is given to the authors, artists, photographers, museums, publishers, and agents for permission to reprint copyrighted material. Every effort has been made to secure the appropriate permission. If any omissions have been made or if corrections are required, please contact the Publisher.

Photographic and Illustrator Credits
Front cover wrap ©Dale Johnson/Aurora Photos
Back cover ©Sofía Ruzo

Acknowledgments and credits continue on page 267.

Printed in the United States of America
Print Number: 02
Print Year: 2019

For product information and technology assistance, contact us at Customer & Sales Support, 888-915-3276
For permission to use material from this text or product, submit all requests online at **www.cengage.com/permissions**
Further permissions questions can be emailed to **permissionrequest@cengage.com**

National Geographic Learning | Cengage
1 N. State Street, Suite 900
Chicago, IL 60602

National Geographic Learning, a Cengage company, is a provider of quality core and supplemental educational materials for the PreK-12, adult education, and ELT markets. Cengage is a leading provider of customized learning solutions with employees residing in nearly 40 different countries and sales in more than 125 countries around the world. Find your local representative at **NGL.Cengage.com/RepFinder.**

Visit National Geographic Learning online at **NGL.Cengage.com/school**

ISBN: 978-13379-11672